AF855693

TOXICOGENOMICS AND PROTEOMICS

NATO Science Series

A series presenting the results of scientific meetings supported under the NATO Science Programme.

The series is published by IOS Press and Kluwer Academic Publishers in conjunction with the NATO Scientific Affairs Division.

Sub-Series

I.	Life and Behavioural Sciences	IOS Press
II.	Mathematics, Physics and Chemistry	Kluwer Academic Publishers
III.	Computer and Systems Sciences	IOS Press
IV.	Earth and Environmental Sciences	Kluwer Academic Publishers
V.	Science and Technology Policy	IOS Press

The NATO Science Series continues the series of books published formerly as the NATO ASI Series.

The NATO Science Programme offers support for collaboration in civil science between scientists of countries of the Euro-Atlantic Partnership Council. The types of scientific meeting generally supported are "Advanced Study Institutes" and "Advanced Research Workshops", although other types of meeting are supported from time to time. The NATO Science Series collects together the results of these meetings. The meetings are co-organized by scientists from NATO countries and scientists from NATO's Partner countries – countries of the CIS and Central and Eastern Europe.

Advanced Study Institutes are high-level tutorial courses offering in-depth study of latest advances in a field.
Advanced Research Workshops are expert meetings aimed at critical assessment of a field, and identification of directions for future action.

As a consequence of the restructuring of the NATO Science Programme in 1999, the NATO Science Series has been re-organized and there are currently five sub-series as noted above. Please consult the following web sites for information on previous volumes published in the series, as well as details of earlier sub-series:

http://www.nato.int/science
http://www.wkap.nl
http://www.iospress.nl
http://www.wtv-books.de/nato_pco.htm

Series I: Life and Behavioural Sciences - Vol. 356 ISSN: 1566-7693

Toxicogenomics and Proteomics

Edited by

James J. Valdes

U.S. Army RDECOM, Edgewood Chemical Biological Center, USA

and

Jennifer W. Sekowski

U.S. Army RDECOM, Edgewood Chemical Biological Center, USA

IOS Press

Amsterdam • Berlin • Oxford • Tokyo • Washington, DC

Published in cooperation with NATO Scientific Affairs Division

Proceedings of the NATO Advanced Research Workshop on
Toxicogenomics and Proteomics
16–20 October 2002
Prague, Czech Republic

© 2004, IOS Press

All rights reserved. No part of this book may be reproduced, stored in a retrieval system, or transmitted, in any form or by any means, without prior written permission from the publisher.

ISBN 1 58603 402 2
Library of Congress Control Number: 2003115502

Publisher
IOS Press
Nieuwe Hemweg 6B
1013 BG Amsterdam
Netherlands
fax: +31 20 620 3419
e-mail: order@iospress.nl

Distributor in the UK and Ireland
IOS Press/Lavis Marketing
73 Lime Walk
Headington
Oxford OX3 7AD
England
fax: +44 1865 75 0079

Distributor in the USA and Canada
IOS Press, Inc.
5795-G Burke Centre Parkway
Burke, VA 22015
USA
fax: +1 703 323 3668
e-mail: iosbooks@iospress.com

LEGAL NOTICE
The publisher is not responsible for the use which might be made of the following information.

PRINTED IN THE NETHERLANDS

Foreword

The field of toxicology has developed a well-characterized set of techniques to assess the behavioral and histopathological consequences of exposure to environmental insults using a number of animal models. These techniques are suitable for determining crude endpoints of exposure such as death, but are not optimal for assessing the more subtle effects of very low level or multi-agent chemical exposures, nor do they offer mechanistic explanations at the molecular level.

More recently, *in vitro* techniques using mammalian cell culture, including human cells, have been developed. These approaches offer high-throughput, inexpensive assays using defined cell types, but are inadequate in situations requiring bioactivation of the toxicant, and do not lend themselves to a systems biological analysis.

We now know that gene activity is exquisitely sensitive to environmental perturbations and that genetic regulation is responsive long before the elaboration of long-term pathologies. It is therefore possible to develop a predictive toxicology based on analyses of the genome, proteome and metabolome.

This book is designed to have a mix of chapters devoted to classical toxicology followed by those focused more on the emerging techniques of toxicogenomics and proteomics. In this way, the relevance of new technologies such as gene arrays to classical toxicologic problems is made evident. Finally, because the worst of the world's toxicology problems reside in developing nations while the latest technical developments are occurring in the industrial nations, we sought to provide a balance of both scientific and geographical perspectives from researchers engaged in toxicology and public health research.

The Editors

Contents

Toxicogenomics and Proteomics
J.J. Valdes and J.W. Sekowski (Eds.)
IOS Press, 2004

Systems Biology

George LAKE
Institute for Systems Biology, 1441 N 34th St., Seattle WA 98103, USA

Abstract. Systems biology uses perturbations (genetic, chemical, biological) and global measurements to interrogate biological systems. The current "sweet spot" for systems biology is the elucidation of "pathways" or "biomodules". Here, the goal is to place genes into their pathways, understand the internal regulation and mechanics of theses pathways and find the interrelationship of the pathways. Ultimately, we would like to construct deterministic models that would enable greater understanding of disease mechanisms as well as understanding the efficacy and multiple (side) effects of small molecules.

Introduction

The Human Genome Project holds the promise of revolutionizing biology. The power of the structure of DNA, first realized by Gamow [1], is that is stores information digitally. This has profound consequences from understanding how inheritance and development can be so faithful to the use of the word "finished" in connection with the scientific project. One can never "finish" other observations of the physical world such as astronomical sky surveys or earth observation. Only a system that is inherently digital can be read and considered finished. It is the program for life, but it comes in a form that is more mysterious than getting the binary of a CAD file for a Boeing 747. The next decades, perhaps millennia of biology will be spent trying to understand the myriad of ways that information has been encoded and flows from DNA thru RNA to proteins to functional proteins to pathways to interacting pathways within a cell to organs and whole organisms. Over the course of time, we should expect many of the abstractions in that chain to be refined and even qualitatively changed. The nature of our abstractions is itself a great challenge.

HGP is also the first of many great "Digital Discovery" projects that will continue to bring revolutionary change to biology. It has begun the global discovery of genes in organisms, the use of cDNA and oligo arrays to globally examine differential gene expression.

1. Digital Discovery

We are all familiar with the rapid pace of information technology, computers double in speed every 18 months, accumulating to a hundred fold increase every decade. At the heart of this increase in power has been VLSI and multiplexing with the use of fiber optics. These same technologies have revolutionized data acquisition systems, detectors and permitted large scale multiplexing of detectors. The result has been a qualitative change in the nature of most sciences, astronomy with Digital Sky projects [2], earth science with the Digital Earth Initiative [3] and biology with the Human Genome Project and the expected follow-ons.

Associated with these changes is a change in the nature of scientific inquiry. Some see the transition from "hypothesis driven science" to "discovery science" [4] as a transition from using accumulated wisdom to fishing expeditions. The main distinction between discovery science and hypothesis driven science is that the former operates on a higher level of abstractions. A typical hypothesis driven researcher may try to examine the role of a single gene/protein in a pathway. In the discovery approach, one might seek to put all genes into pathways and explore their interconnections. Discovery science is often a matter of "finding targets" (in a broader sense than "drug targets") where detailed biochemistry will likely determine the exact role of an identified target. I note that the arguments around these differences are common to all the fields being impacted. In astronomy, arguments abound over the value of "studying one star at a time" versus a global survey of the sky. With the substitution of very few words, one can create statements often heard in the halls of biology institutes and meetings.

One difficulty in operating on higher level abstractions is that the abstractions themselves are still evolving with the theoretical underpinning of biology. At the moment, the "sweet spot" for systems biology is pathways and networks [5]. These abstractions have evolved over many decades. They are accessible by global means and clearly connect to medical research. However, we fully expect that the meaning of these networks will evolve considerably over the next 20 years. For now, we trying to understand how information goes through signaling pathways, how gene expression is regulated, how protein production and post-transcriptional modifications are triggered.

Over the last few years, it has become increasing clear that the nature of gene networks and pathways is that they are highly modular and hierarchical. With this comes the ability to employ new abstractions where we can make some headway by separating the goals of determining which genes are within a module as well as the main inputs and outputs. In this way, we roll up a large of amount of information to achieve a more global view. The dynamics of a network can be described by all of the detailed (stochastic) rate equations in the same way that fluid dynamics must result from the interactions of all the molecules within the fluid. However, a formulation of the fluid equations in the form of Navier-Stokes makes no reference to the interactions of the molecules and is far more appropriate for the design of aircraft. Similarly, I suspect that within a few years, we will be using systems of equations to describe the function, perturbation, failure and repair of networks that make no specific reference to the chemical rate equations. But, the first step in this task is certainly to identify the complete genetic parts list for module and develop some understanding of the structure of the network.

One likely shift will be from an emphasis on the genes to a greater emphasis on control structures. The past emphasis has been on the DNA itself as a "program for life". During the beginning of the explosion of the information age, we saw a paradigm shift from life as "machine" to life as "computer". As the field has matured and we've entered the internet age, we are coming to view life as an interacting network of computers. Similarly, the view of the "program of life" as a long list of statements is giving way to the idea of modular and hierarchical abstractions. That is, way that we see life coded has shifted from a long list of assembly statements to a more "object-oriented" approach. This approach is taken by programmers to build solutions that can be maintained and evolve. We seem to see the same approach within biological systems. The way that computer code is recycled in an evolutionary approach is to alter the control statements. In this way, a fluid module might be used in a climate model or a simulation of an aircraft. We might expect to see a similar approach to recycling the use of genes during evolution.

At the present time, we know that there is an extremely high degree of "code reuse" in biological systems. A dramatic example is the 92% sequence similarity between human and mouse genomes. Yet, 92% of the drugs developed in mice certainly don't work in humans, a mouse isn't 92% of the size of humans, so the true meaning of this similarity figure is elusive. Genes themselves are fairly large, but control regions tend to be a smaller number of bases. As a result, we might expect a more rapid evolution in control structure than in the program modules. Understanding this evolution of control structures is likely to be the most revolutionary change in biology during the next 20 years.

2. Global Experiments

There are a variety of global techniques that can now be applied to biological systems:

Determination and comparison of genomes
Global expression profiling with arrays
Measuring the Proteome.
Determining the Interactome

3. Computation and System Biology

George Gamow pointed out that the singular property of DNA is that it means that biological information is inherently digital thus providing a solution for a variety of problems including the fidelity of inheritance. This property of DNA is unique in science, I know of no other case where the observation of the physical world captures a system that is inherently digital. As a result, the language of the science of sequencing also has remarkable properties. We routinely talk of "finishing the human genome". You can't hear anyone say that they are going to finish surveying the sky or colliding particles in accelerators.

Sequencing and analyzing DNA has been one of the main ways that computation has entered the world of biologists. While we still read in Time magazine that it takes a box of floppy disks to hold the genome (indeed, this is the only time that we read about floppy disks!), the genome is really rather small fitting easily on a CD ROM. It is also one dimensional and static with only modest variations between individuals. Nonetheless, extracting all the information of multiple genomes is a sizeable problem, the scope of which is not even completely understood today.

If biology stopped at DNA, we'd have a challenge, but nothing compared to that of dealing with the multiple levels of information that follow. At this point we enter a world with multiple branchings, such as the alternative splicing of genes and wealth of post transcriptional modifications of proteins. Will the small numbers of transcripts per cell, stochastics become important. The data varies with the time and environment and usually has properties that are more analog than digital. We are able to capture this data at a phenomenal rate.

We can learn a number of features of high volume data from other fields such as astronomy and particle physics. These features include:

Technology can break solutions. We know that the power of computing doubles every 18 months, integrating to a factor of 100 each decade. This is fantastic news if you are Walmart with a transaction rate that increases at 15% a year, sufficient to be a Wall Street darling. Even if you look at all the relationships among the data, this increases 30% a year. With the growth in computing power, this gets 40% easier every year. You can think of more and more Byzantine things to do, even though the reported discover of a correlation of "beer" with "diapers" is an urban myth, there must be such peculiarities to be discovered. In contrast, scientific detectors are built from VLSI and are often connected by fiber optics—exactly the same technology that causes this explosive growth of computing power. Hence, scientific data rates pace the growth in computing power. If one does anything that scales more than linearly with the data rate, technology will break a problem that is currently solved.

Data is collected that no scientist ever examines, it processed by machines and viewed only statistically. With greater automation of data collection and data rates that are growing at the rate of Moore's Law, data is collected that is only examined as part of a distribution with the focus on "outliers". As a result, we are increasingly reliant on the filters that we build into analysis programs. There are approaches to trying to find "things of interest" without specifying too much what that means. These will be increasingly important, but for now, we will find things that we expect to find and the rest may stay hidden for sometime to come.

Statistical inference takes some subtle shifts. Normally, if you measure one quantity, you are likely to happy with a significance value of $p = 0.01$ or perhaps even $p = 0.5$. However, if I use a microarray with 10,000 spots, $p = 0.05$ will deliver roughly 500 spots that are bogus. As a result, one winds up setting p = 0.5/(the number of spots) so that it becomes slightly unlikely that even one bogus spot is in the sample. Although, the error distributions often have "heavy tails", so one should never take the idea that "all the selected data is significant" too seriously. This has lead to notions that techniques that only measure one variable, like a Northern blot, are more sensitive than high throughput techniques like arrays. This is more a difference in what question is asked. If you suspect a linkage of a particular gene to a disease or process, then seeing that gene change with a $p = 0.01$ on an array of 10,000 genes is a significant result. However, if you are in the discovery mode and use that threshold of $p = 0.01$ to fish for genes that are linked, then you'll get ~100 bogus ones.

Persistence of data is important, new data makes old data more valuable. Global data becomes ever more important. Each new genome makes comparative genomics more powerful. The genomes are the prerequisite for transcriptomics and proteomics. Weak trends will only emerge by being able to examine large datasets.

With the break between humans and the data, scientists must have control of the tools, it's not something that can be contracted. Scientists must have active control of their tools. The current approach in biology is slowing progress and will eventually hit a wall. This is a lesson being painfully learned by the climate community which is a perennial focus of Government Accounting Office Reports for just this reason.

Let me illustrate some of these points with the example of the extremely large Sloan Digital Sky Survey (SDSS) [2]. This survey looks at 1/4 of the entire sky with 5 different filters which astronomers call "colors". There are roughly a trillion pixels or independent spots given the survey resolution. At the survey depth, it will detect roughly 1 Million stars and 10 Million galaxies. In any given "color", the goal is that only 5% of the sources will be bogus, which means that we want to keep the bogus count down to roughly 500,000. With one trillion pixels, this requires $p = 0.0000005$ or roughly 5 standard deviations (5σ). With this significance cut, most sources will have counterparts in the other colors (some very interesting sources like quasars may not), so while each color is 5% bogus, the merged total catalog is 20-25% bogus.

If I add new data, then the value of "surveys" increases. In the sky survey example, if I detect a source in the sky by some other means, let's suppose that it flashes X-rays, then I can go back to my SDSS data and look for a source that might be just a 3σ detection, because I'm examining one spot not a trillion.

4. Illustration of the Global Approach

There are a variety of global methods emerging, but the most available is the interrogation of gene expression using cDNA arrays. The Rosetta compendium [6] illustrated the power of this technique for functional genomics. In this study, they deleted genes that had not been previously characterized. They were able to infer function and characterize the unknown target of a common drug by matching the subsequent expression profiles to those of deletions with known function. A group at ISB [7] performed a systematic study of galactose metabolism in yeast. In addition to cDNA array data, they added available data on protein expression as well as data on protein-protein and protein-DNA interactions. They ran strains with each of the 9 genes in the galactose network deleted. They saw significant changes in the expression level of 997 genes (our of a total of ~6200). The expression profiles clustered to enable the classification of genes in metabolic, cellular and synthetic pathways. They built a model of the regulatory network that revealed that a particular double deletion would be a sensitive probe of the model's uncertainties, showing that their model was a useful guide to further experimental design.

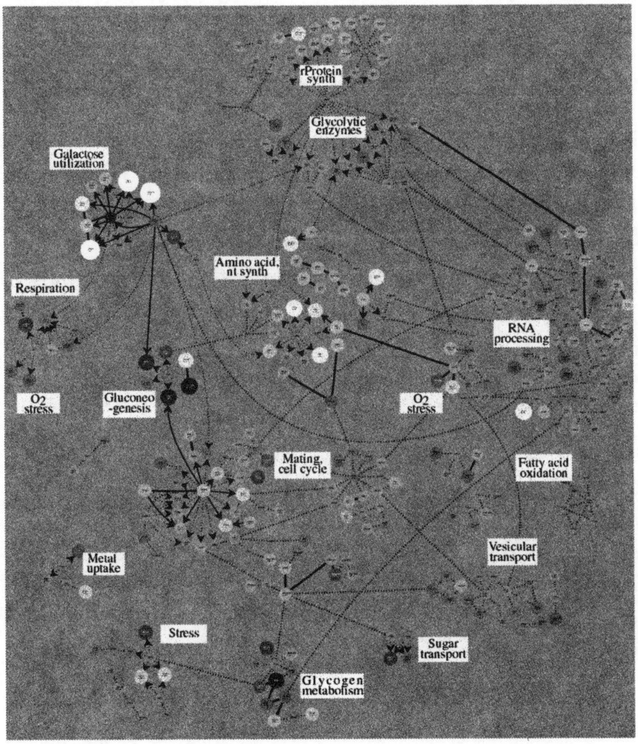

Figure 1: Cellular networks in yeast revealed by pertubing the galactose network [7]. Each circle represents a gene observed to vary on the cDNA array. The size and color depicts the observed variance of the expression level together with its significance under the deletion of *gal4* and availability of galactose. Directed arrows indicate protein-DNA interactions, undirected lines are protein-protein interactions. Figure courtesy of Ideker, Thorsson, Hood *et al.* [7]

The most interesting global study that I've seen to date is that of Rives and Galitski [8]. They developed an unsupervised technique for revealing the hierachical and modular structure of yeast from data on the interactome, the network of protein-protein interactions

[9,10]. Schwikowski and collaborators [10] had shown that interaction partners are an effective way of predicting the function of proteins. This uses the most local of information in the interactome. Rives and Galatksi wanted to look at structure on somewhat larger scales to understand how the proteins are related to one another in modules and connections between modules. To do this, they took the symmetric matrix that describes the structure of the protein network, each (n,m) entry is the distance of the shortest path between proteins. They then construct a "profile matrix" where each entry is a "weight" divided by the square of the above entry. This takes a global look at the interactions, but progressively distinguishes less between distances as they become larger. The "weight" can be some other data of interest, such as the correlation between expression changes in a sequence of experiments. Each column (or row since it's symmetric) represents a global interaction profile for a single protein. These vectors are clustered and the matrix reveals a network of interconnected modules. A subset of the data is shown in Figure 2, where the different modules have been characterized with their known functions.

Creative use of "weights" holds considerable promise and illustrates how new value comes to old data by virtue of additional data and analysis. There are a variety of competing and cooperating databases to achieve this cumulative power of data. There is great hope for array data now that standards are emerging, but one is hard pressed to "find the data" even when journals nominally require it to be public as a condition of publication. Nonetheless, one can explore a number of archives of array data [12-16], protein interaction data [17-19], protein-DNA interaction data [20-21] and metabolic pathway data [22-24]. Nat Goodman [25] does periodic reality checks and updates on available data in his column for Genome Technology.

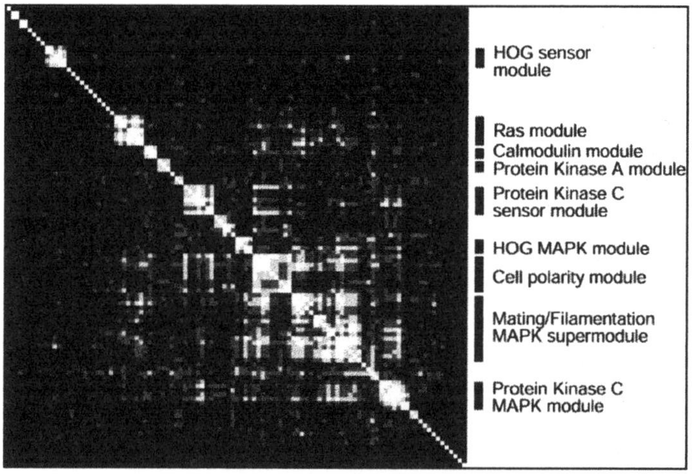

Figure 2: The technique of Rives and Galitski applied to the signal-transduction module network of yeast. One can see the clear modular structure as well as the connections between modules. The modules were found in an unsupervised fashion but can be identified as known pathways. Figure courtesy of Tim Galitski.

5. Summary

In science, practice always follows theory. Strangely, theory has a bad name in biology. It seems to be confused with "wild hypothesis" as I routinely hear the dismal "that's only a theory". But, theory is absolutely the best thing that science has to offer, quantum mechanics are relativity are theories. That said, models that aren't well tied to a theories are often junk. In the next decades, I think we will see theory emerge in biology as strongly as it has in physics.

This remarkable transition to a science with a theoretical underpinning that will enable biologists to take a true engineering approach to applications as diverse as medicine and organic sensors. At the moment, the basic theory is evolution that tells us to try a million things and see what happens. Drug companies pride themselves on the size of their chemical libraries; the largest pharmas can literally follow theory and "try a million things and see what happens". It seems almost like structural engineering in the middle ages, if the cathedral fell, the next version was given another flying buttress.

The current danger is that while a million things can be tried in a laboratory before the release of a new drug therapy, the general population tries billions of things with their complex genetic makeups and drug interactions. As a result, we've seen a couple drugs per year withdrawn after approval owing to current complexity of multiple drug interactions. With various sorts of high throughput experiments, we will see greater fingerprinting and characterization of drugs allowing us to zero on possible "interactors". Again, this highlights the need for persistent information to enable the comparison of drug profiles.

I would like to make one last comparison between biology and the space science culture that I was raised in. I often hear that the Human Genome Project has revolutionized biology. If that's true, where's the next project of this scale? NIH could do a few projects of this size every year using only their budget increases, yet not one such project has been started. In contrast, space physics is small, but they start two or three projects a decade that cost over $2B; the Hubble Space Telescope is one of four great observatories that was launched within a decade. Part of this is probably the conservatism that arises from a tight coupling to medicine. I don't think it's any accident that HGP was started by DOE rather than NIH. One of my hopes is that we'll see larger scale projects undertaken as the result of greater interest in biology by DOE, DARPA, NATO and Homeland Defense's Office of Research.

References

[1] G, Gamow, Possible relation between deoxyribonucleic acid and protein structures, *Nature* **173** (1954) 318-319.

[2] A. S. Szalay, P. Z. Kunszt, A. Thakar, J. Gray, D. R. Slutz, The Sloan Digital Sky Survey: Designing and Mining Multi-Terabyte Astronomy Archives, *SIGMOD Conference 2000*, 451-462

[3] The Digital Earth Initiative Consortium, http://digitalearth.gsfc.nasa.gov

[4] R. Aebersold, L. Hood. and J. Watts, Equipping scientists for the new biology, *Nat. Biotechnol.*, **18** (2000), 359-360.

[5] L. H. Hartwell, J, J. Hopfield, S. Leibler and A. W. MURRAY, From molecular to modular cell biology, *Nature* **402** (1999) C47 - C52.

[6] T. R. Hughes, M. J. Marton, A. R. Jones, C. J. Roberts, R. Stoughton, C. D. Armour, H. A. Bennett, E. Coffey, H. Dai, Y. D. He, M. J. Kidd, A. M. King, M. R. Meyer, D. Slade, P. Y. Lum, S. B. Stepaniants, D. D. Shoemaker, D. Gachotte, K. Chakraburtty, J. Simon, M. Bard, and S. H. Friend.\, Functional Discovery via a Compendium of Expression Profiles, *Cell* **102**, (2000).109-126.

[7] T. Ideker, V. Thorsson, J. Ranish, R. Christmas, J. Buheler J. K. Eng, R. Bumgarner, D. R. Goodlett, R. Aebersold, L. Hood, *Science* **292** (2001) 929-934.

[8] A. W. Rives and T. Galitski, Modular Organization of Cellular Networks, PNAS, **109** (2002), 1128-1133.

[9] P. Uetz, L. Giot, G. Cagney, T. Mansfield, R, Judson, J. Knight, D. Kockshon, V. Narayan, M. Srinivasan, P. Pochart, *et al. Nature*, **403** (2000), 623-627.

[10] T. Ito, T. Chibaa, R. Ozawa, M. Yoshida, M. Hattori, and Y. Sakai, *PNAS* **98** (2001), 4569-4574.

[11] B. Schwikowski, P. Uetz and S. Fields, Nat. Biotechnol. 18 (2000), 1257-1261.

[12] J. Aach, W. Rindone, G. M. Church, Systematic Management and analysis of yeast gene expression data, *Genome Res.*, **10** (2000) 431-445.

[13] O. Ernolaeva, M. Rastogi, K.D. Pruit, G.D. Schuler, M.L. Bittner, et al. Data management and analysis for gene expression data, *Nat. Genet.* **20** (1998) 19-23.

[14] V. Hawkins, D. Doll, R. Bumgarner, T. Smith, C. Abajian et al. PEDB: the Prostate Expression Database, *Nucleic Acids Res*., **27** (1998) 204-208.
[15] M. Ringwal, J. T. Eppig, J. A. Kadin, and J. E. Richardson, GXD: a Gene Expression Database for the laboratory mouse: current state and recent enhancements. The Gene Expression Database Group, *Nucleic Acids Res*., 28 (2000), 115-119.
[16] C.J. Soteckert Jr., F. Salas, B. Brunck and G. C. Overton, EpoDB: a prototype database for the analysis of genese expressed during vertebrate erythropoiesis. *Nucleic Acids Res*., **27** (2000) 200-203.
[17] I. Xenarios, E. Fernandez, L. Salwinski, X. J. Duan, M.J. Thompson, et al, DIP: the database of interacting proteins. *Update, Nucleic Acids Res*., **29** (2001), 239-241.
[18] G. D. Bader, I. Donaldson, C. Wolting, B. F. Ouellete, T. Pawson, et al, BIND—The biomolecular interaction network database, *Nucleic Acids Res*., **29** (2001), 242-245.
[19] H. W. Mewes, K. Heumann, A. Kaps, K. Mayer, F. Pfeiffer et al. MIPS: a database for genomes and protein Sequences, *Nucleic Acids Res*., **27** (2000), 44-48.
[20] E. Wingender, X. Chen, R. Hehl, H. Karas, I. Liebich et al, TRANSFAC, an integrated system for gene expression regulation, *Nucleic Acids Res*., **28** (2000), 316-319.
[21] J. Zhu and M. Q. Zhang, SCPD: a promoter database of the yeast Saccharomyces cerevisae, *Bioinformatics,* **15** (1999) 607-611.
[22] P.D. Karp, M. Riley, M Saier, I. T. Paulsen, S. M. Paley et al, The EcoCye and MetaCye databases, *Nucleic Acids Res*., **28** (2000), 56-59.
[23] H. Ogata, S. Goto, K. Sato, W. Fujibuchi, H. Bono et al, KEGG: Kyoto Encyclopedia of Genes and Genomes, *Nucleic Acids Res*., **27** (2000), 29-34.
[24] E. Selkov Jr., Y. Grechkin, N. Mikhailova and E. Selkov, MPW: the Metabolic Pathways Datavase, *Nucleic Acids Res*., **26** (1998), 43-45
[25] N. Goodman, A Plethora of Protein Data, a Shortage of Solutions, *Genome Tech.,* **22** (2002) 82-88.

Toxicogenomics and Proteomics
J.J. Valdes and J.W. Sekowski (Eds.)
IOS Press, 2004

The Role of Bioinformatics in Toxicogenomics and Proteomics

Bruno SOBRAL[1], Dana ECKART, Reinhard LAUBENBACHER and Pedro MENDES
Virginia Bioinformatics Institute
Virginia Polytechnic Institute and State University
1880 Pratt Drive (0477), Blacksburg, VA, US

Abstract. Modern biology is now challenged with understanding living organisms on a systems level, empowered by the enormous success of molecular biology and numerous genome projects. Bioinformatics provides a research platform to acquire, manage, analyze, and display large amounts of data, which will in turn catalyze a systems approach to understanding biological organisms as well as make useful predictions about their behavior in response to environmental and other perturbations. This task, to mathematically model living organisms, currently drives computational infrastructure development, computer modeling and simulation, and the development of novel mathematical approaches. Infectious disease research is one area of research that now requires the development of robust information technology infrastructure to deal with vertical and horizontal data integration. It will also benefit greatly from novel biochemical and mathematical modeling and simulation approaches. Significant advances in computational infrastructure, for example, the use of Grids and web-based portals, will likely contribute to the decompartmentalization of scientists and funding agencies, which currently hinders the utilization of infectious disease data. PathPort, short for Pathogen Portal, offers an example of a collaborative software development effort designed to develop, test, and implement a robust infrastructure for data management, integration, visualization, and sharing. PathPort's main research goal is to build a scalable portal providing data management and analysis for key pathosystems, especially relating to molecular data. Integrated molecular data sets (e.g., genomics, metabolomics, transcriptomics, and proteomics) will provide workflows that assist with development of novel diagnostics and countermeasures to combat infectious diseases. The study of networks, both via novel simulation and mathematical modeling approaches, will also catalyze our comprehensive understanding of biological systems.

1. Introduction

The fundamental goal of biological research is to understand life at various levels of organization, such as genes, proteins, chemicals, and cells. It is of interest to biologists to comprehend how the lower levels of biological organization also provide the basis for understanding higher levels of explanation, such as physiology, anatomy, behavior, ecology, and populations. A paradigm shift in biological research philosophy, notably from a reductionist approach to an integrative approach seems to be occurring recently.

The last century might eventually become known in biological research as the "century of the gene". In the 1900s, we witnessed phenomenal success in understanding the nature of inheritance (genetics) through the discovery of deoxyribonucleic acid (DNA) and the subsequent emergence and success of molecular genetics. These culminated in the many genome projects that were successfully undertaken. In the century of the gene, biology embraced a reductionist approach resulting in an increased molecular understanding of components of living systems. Today, we struggle to understand the whole organisms using our recently acquired molecular information.

The 21st century may very well become known as the "century of the biological system". The enormous success of molecular biology and the engineering of high-throughput data production systems for the analysis of biological molecules [DNA, ribonucleic acid (RNA), proteins, and metabolites] have caused this designation in part. The resulting data have spawned the need for computerized data management to become a central infrastructural component of biological research. One component of which is currently called bioinformatics. However, it is not only the explosive growth of data that pushes us into the biological system. The need to synthesize the molecular data into unified concepts that help us understand how cells, tissues, organs, organisms, populations, habitats, communities, and ecosystems work also serves as a catalyst. In a sense, the particle physics community has been struggling with a similar problem—the derivation of a unified theory from an increasing understanding of the components.

Bioinformatics has many definitions since it has become a central component of genome projects. Early use of the word during the last century by the theoretical biochemistry community pointed towards a mathematical understanding of living organisms and their processes [1]. More recently, genomics-inspired definitions of bioinformatics tend to be related to the use of information technologies to acquire, store, analyze, and display large amounts of molecular data, especially DNA sequence data [2]. A recent article in *The Economist* defines bioinformatics as "a branch of computing concerned with the acquisition, storage and analysis of biological data." Moreover, it speaks of bioinformatics as a spectrum of technologies, covering such things as computer architecture, storage and data-management systems, knowledge management and collaboration tools, and the life-science equipment needed to handle biological samples [3].

As a result of genome projects and other high-throughput biological analyses, biological data are growing faster than the rate at which one can obtain twice the central processing unit (CPU) power for the same money (known as Moore's Law). Because of this, the computational cost of doing a similarity search at a public database such as GenBank has increased over the last ten years, despite the increasing performance of CPUs as empirically shown by Moore's Law. The Information Technology (IT) industry, especially hardware manufacturers, has embraced this situation because of potential sales. In addition, as our molecular understanding of biological systems has grown, algorithms—hardware and software—have in some cases mimicked biological systems to provide new technologies and approaches. Thus, we frequently hear of "convergence" and see emergence of DNA computing [4], genetic algorithms [5], [6], and software development in an object-oriented world that feels very biological. DNA computing is interesting because it is not just a mimicry of a biological system; it actually uses real DNA to compute. We may be witnessing the replacement of physical sciences with biological sciences as the main driver for the development of information technologies. IBM's Blue Gene project serves as concrete example of this [7].

Systems science grew out of a series of multidisciplinary meetings held by Wiener, von Bertalanffy, Asby, and von Foerster in the 1940s and 1950s [8]. The central concept (of what was then called cybernetics and systems science) was that regardless of the world's complexity and diversity, some concepts and principles would always be domain independent. Wiener and others held that uncovering these general laws would allow us to apply them to solve problems in any domain. Systems science differs from the more analytical approach in that the emphasis of systems science is on interactions and connectedness of the components in the system. In biological systems, Wiener used the term cybernetic to state that these were complex, adaptive, self-regulating systems [8].

As biological research grew from a cottage industry science to "big science" in the last century, a number of new features evolved. First, biological research infrastructure became increasingly expensive. Lots of people needed to use the new high-performance infrastructure, leading to challenges in design, implementation, and management of such infrastructure, especially by cottage industry biological researchers. Second, a clear need for integration across disciplines, especially those outside of what was traditionally considered biology, emerged. Third, experimentalists, who have long had the upper hand in biological research, were confronted with the need to work closely with theoreticians as biology grew from data-poor and theory-rich to data-rich and theory-limited. Again, parallels with the physics community's path emerge.

2. The Need for Data Integration: An Example from Infectious Disease Research

2.1 Background

Biologists now have the capabilities to develop and access very large, genome-scale data sets for a number of organisms. As of late 2002, approximately 90 complete genomes are in the public domain and over 120 more are in progress [9]. Even larger data sets for transcriptomes, proteomes, and metabolomes are being developed concurrently. Aside from management challenges presented by this tremendous amount of data, many would like to use it to understand living systems and make useful predictions about their properties or behavior in response to perturbations, such as specific environments. To effectively use molecular data to make predictions about biological processes, the data must first be accessed (Figure 1). It is also necessary to query and analyze the data in the context of the environmental conditions of interest. And, in many cases the environmental or contextual data is not acquired or stored as metadata. Taken together, this situation calls for data management that enables data integration horizontally (within a data type, e.g., DNA sequences and across organisms) as well as vertically (across data types, e.g., DNA sequences with gene, protein, and metabolite expression profiles).

Many completely sequenced genomes are prokaryotic [9]. Many of these sequenced organisms are pathogens of humans, animals, and food crops. Infectious diseases pose a major ongoing global problem that spans agriculture, environmental sciences, veterinary medicine, and biomedicine. The interaction between hosts, pathogens, and the environment is frequently referred to as the "disease triangle" in plant pathology. Multiple data types could be acquired and developed if we consider the integrated "host-pathosystem" or "disease triangle" and applied advanced laboratory technologies to the problem (Figure 2). Therefore, infectious disease research is a reasonable starting place to develop robust information technology infrastructure for dealing with vertical and horizontal data integration. Not only is there high data density, the applicability of the results has significant ramifications. The testable hypothesis is whether or not comprehensive data integration is feasible; and, if it can be achieved, whether or not the resulting integrated data sets make it easier to extract knowledge, meaning, and utility from the data.

Comprehensive data integration poses significant challenges for the broader scientific community. Biological researchers have been and continue to be fragmented into disciplinary and application dimensions. For example, scientists working with human infectious disease tend to be in medical schools, while scientists working with plant infectious diseases tend to be in agricultural colleges. Worse, in many cases the specialization is such that the pathologist works on the microbial pathogen while the host specialist is in another department thereby fragmenting the study of the pathosystem. A fragmentation in funding agencies that support infectious disease research also exists. Typically, scientists write to the United States Department of Agriculture (USDA) for support if they are studying plant pathosystems and to the National Institutes of Health (NIH) if they are working on human pathosystems. More than 40 federal agencies are interested in funding or using results of infectious disease research in the United States. The combined compartmentalization of scientists and funding agencies results in ineffective utilization of resources and data [10].

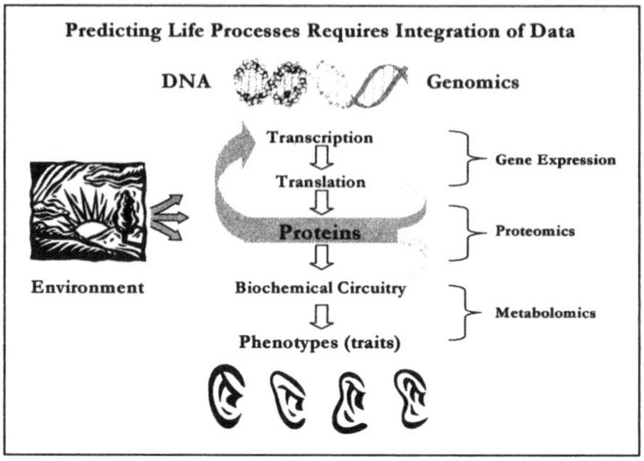

Figure 1. Simplified representation of key life processes and data types that can be extracted in massive parallel analyses of molecular data.

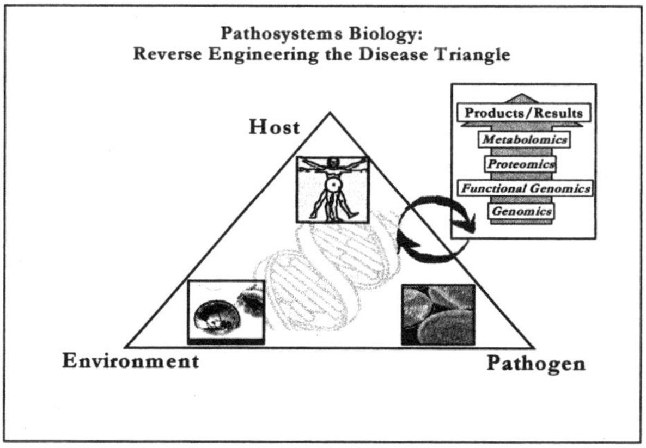

Figure 2. A systems view of hosts, pathogens, and environmental factors that contribute to results of encounters between host and pathogen organisms in field conditions, and infectious disease research data types that may be influenced.

Pathogens can establish themselves in regions of the world by three major routes: 1) accidental introduction through movement of people and goods, 2) natural evolution of local pathogens that have been kept in check through environmental or chemical controls, and 3) intentional introduction. The last of these has recently received significant media attention. Regardless of intentional events, pathogens will always put pressure on their hosts through parasitization. In other words, the "arms race" between hosts and pathogens in their struggle for survival will sustain the need for infectious disease research indefinitely. This co-evolution suggests that we will never be entirely rid of infectious diseases. Even when a particular pathogen is excluded from a particular host, another pathogen will leverage that competitive void to infect that host.

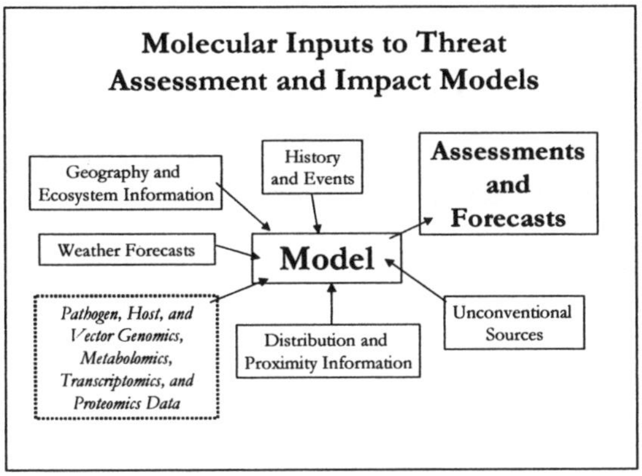

Figure 3. Assessment and input models currently use many inputs. To date, integration of comprehensive molecular inputs has been missing, represented here by the box at lower left.

Molecular toxicology studies focus on cellular responses to chemicals. Pathogens cause some effects in hosts due to chemicals they produce, in some cases called toxins. Therefore, similar approaches can be leveraged across infectious disease and toxicogenomics research agendas. For example, cellular response to chemicals is central to both areas of investigation. By using molecular data acquired through modern laboratory technologies in combination with environmental data, we can model outcomes of host and pathogen encounters in field conditions (Figure 3). In addition, significant data sets, especially at the DNA level (genomes), should be integrated and leveraged further.

2.2 Pathogen Portal (PathPort): A Data Integration Infrastructure Based on Web Services

PathPort is a collaborative software development effort to develop, test, and implement a robust infrastructure for data management, integration, visualization and sharing [11]. PathPort's main research goal is to build a scalable portal providing data-management and analysis for key pathosystems, especially relating to molecular data (DNA, RNA, proteins, and metabolites). The integrated molecular data sets will be useful in, among other things, providing workflows that assist with development of novel diagnostics and countermeasures (vaccines and therapeutics) to combat infectious diseases. Molecular data sets, when combined with ecosystems level data (weather data, for example) will help the construction of molecular epidemiological maps of global proportions, assisting in disease surveillance. As a result of its data management infrastructure, PathPort will assist geographically distributed experts to increase the fundamental understanding about pathosystems.

2.3 The PathPort Information System

ToolBus, a client-side interconnect built on a "bus" architecture, allows the construction of large systems from numerous small and simple modules by extending the client-server model so that both sides are collections of multiple communicating entities, rather than single programs (Figure 4). XML communication between the client and server sides promotes openness and maintains separateness. Toolbus' architecture allows data from different data model instances to be grouped with one another. On the server-side, Web-services are used to yield processed information to the client-side. This system provides an open and scalable means for specifying and accomplishing (bio)informatics processing.

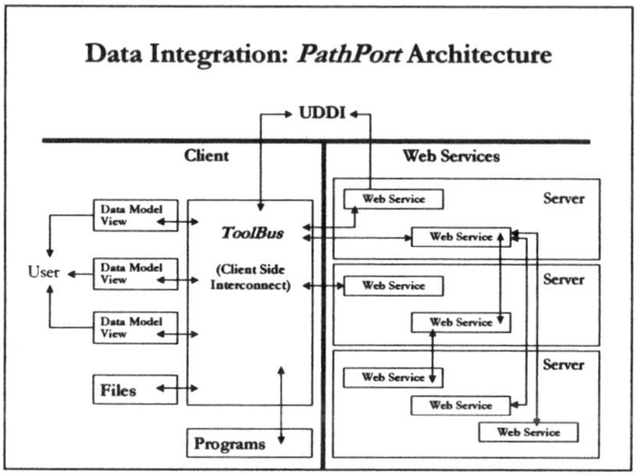

Figure 4. *ToolBus* connects server data sources and analysis tools with client visualization plugins. UDDI refers to Universal Description, Discovery and Integration.

PathPort utilizes Web-services and XML to support remote data access and analysis while relying on plug-ins for information visualization. Other approaches have been tried and all of them, including PathPort's, have been developed to facilitate the data-analysis-visualization workflow, relying on centralizing one or more aspects of the workflow.

Grid systems and *Web-based portals* centralize both data and analysis, whereas larger desktop applications combine analysis and visualization. Grids enable data and computer resource sharing across a diverse set of computer systems and organizations while presenting the end user with a single unified virtual computer system. Resources within a Grid are made available to members of a virtual organization. Virtual organizations correspond to the virtual system that members of that organization see. Avaki [12], a commercial product grown out of the Legion effort [13], and Globus [14], an Open Source project, are the two primary Grid architectures currently available.

Ideally, a user could make use of data from databases residing on seven different computer systems housed within five separate organizations on three continents, for example. Likewise, analysis programs are available only on a small subset of system architectures. The Grid will move the data to those systems able to run the desired analyses, automatically migrating data from databases to those systems running the analyses. The results of one analysis would similarly be migrated to other parts of the Grid where subsequent analysis would take place. The Grid user sees a large diverse collection of data and analysis tools without knowing on which systems the data physically reside or on what architectural platform different analysis programs must be run. If the researcher's desktop computer is part of the Grid, visualization of analysis results can be done directly on the desktop; otherwise visualization can be accomplished by remote display via X11, a static Web-based image, an interactive Web-based Java applet, or similar mechanism, or by downloading the results and using a desktop application independent of the Grid.

If the virtual organization to which a researcher belongs has access to needed data and analysis and visualization tools, no difficulties arise. However, it is common for an individual researcher, or a research group, to have collaborations with a number of other individuals and groups, who may be unwilling to participate in the same virtual organization. In this case, the data and tools available from such collaborations probably will not be available on the Grid. Furthermore due to concerns over data and information control (e.g., with respect to first publications), some collaborators might be unwilling to allow their data and information to be uploaded into a Grid for fear that it could be accessed by others in the virtual organization who are not their collaborators. In fact, primary researchers may be unwilling to upload data into a Grid for the same reason. While the advantages of Grids are clearly evident, the disconnects between different virtual organizations or Grids are limiting factors for many researchers.

Web-based portals as front ends to a server solution are exemplified by BioMed Grid Portal [15], BioASP [16], Technical Computing Portal [17], Zerosum [18] and many others. These solutions almost always combine the database and analysis parts of the workflow into a single system. Increasingly, visualization support is made available either as dynamically generated images or interactively as Java applets. S-Plus, [19] a statistical scripting language with visualization support built on top of Java components, is designed for this purpose. These portals can be used as a front-end to a Grid, alleviating the command line interface problem. Portals suffer from the same issues as Grids. Typically built around a single database and supporting a limited set of tools, they are only able to run on the single platform that often underlies such systems [17], [18].

In response to the single platform limitation, some portals use *portlets* to allow access to different systems [20], [21]. Portlets are independent components that may originate from different sources and are presented within the same Web page. This enables a wide variety of information from different sources to be made available in an end user-configurable manner. If different systems hosting portlets are part of the same virtual organization within a Grid, then the results of a database lookup or analysis tool would be available to other analysis tools. However as seen with Grids, if the systems are not part of the same Grid virtual organization, the user is left hanging. At a minimum, users must download from one Grid to another in addition to maintaining multiple logins and mentally switching from one system to another. Moreover, many Web-based portals do not allow end users to upload their own analysis tools thus forcing users to download, analyze, and upload before continuing with the remainder of their analysis workflow. As seen with Grids, some researchers may be unwilling to upload their data to a portal in order to make use of its analysis and visualization tools for fear of losing control over its use. Concerns about data and information security are of particular interest to both researchers hoping to be the first to publish as well as companies wanting to be the first to market.

Another approach is to centralize the analysis and visualization components of the workflow. These might be called *super applications* both because the number of analysis routines these applications might need to include may be numerous and because at least some of those analyses are likely to require substantial compute resources. J-Express Pro is an example of a super application [22]. Super applications do not address the data acquisition problem and generally expect the data to be contained in local files. In addition, super applications do not allow plug-ins for new analysis or visualization components. This may limit researchers to those tools that the super application developers deem sufficient.

PathPort's approach is to have the client-side serve as the focal point with data and information brought to the desktop for visualization, but off-load data access and analysis to the server-side. Thus, as with super applications, there is responsive interaction for information visualization. But unlike Grids and portals, ToolBus is not limited to a single server or (virtual) organization for acquiring data and accessing analysis tools. Instead, ToolBus relies primarily on Web-services for obtaining both data and analysis, giving it the ability to cut across both machine and organizational demarcations. Web-services are expected to return results in XML format which ToolBus uses to determine the appropriate plug-in(s) to use for visualization. In this way, ToolBus leverages Web-services to enable access, either directly or indirectly, to multiple Grids and portals while utilizing standardized XML formats to ensure interoperability between the various data and analysis sources and automatic visualization tool determination. With separate plug-ins forming the basis for ToolBus' information visualization, the architecture incorporates a mechanism for creating groups of related information taken from one or more visualization tools. Groupings can then be compared individually or in combination to find both common and unique elements using the set operations' union, intersection, and complementation by constructing Venn diagrams.

ToolBus has been designed and developed as a client-side interconnect, allowing data from a variety of sources (local files and programs, Web-services, and other unknown tools) to be connected with a variety of "visualizer" plug-ins. In turn, data from the *Views* of a *Model* can be dragged-and-dropped into *Tools*, forming a feedback connection (an ability typically missing from Web-based portals), or into the *Views* of other *Models*, allowing interactions between separately designed and developed plug-ins. Finally, the *Associator* and the *Group* comparison facility that ToolBus supports allow connections between separately designed and developed *Models* to be made. One early benefit from the architecture was that members of the PathPort development group could easily be divided into three separate teams, working on ToolBus, plug-ins, and Web-services with relatively little interaction required between them. Communication between Web-service and plug-in developers only required agreement on the XML format to utilize. Because of standardization of HTTP, SOAP, and WSDL protocols [23], essentially no communication was required between Web-service and ToolBus developers. Our decision to pass XML document results as Strings, as opposed to more complex data types, greatly simplified the interaction between these two development teams. This division of labor allowed nearly complete independent development of the primary components required by the PathPort project.

In addition to allowing a high degree of task separability, PathPort uses a simplified spiral software development model [24] that enables user participatory design and provides the project with a great deal of flexibility since each of the three elements of PathPort (ToolBus, Web-services, and plug-ins) may be independently designed and developed using the spiral model. The incremental nature of spiral development forces developers to frequently have functional versions from which user feedback can be elicited.

The unique combination of features contained within ToolBus enables a rich set of capabilities for supporting rapid, scalable, and platform independent development of large systems, such as PathPort, which span a wide variety of information types within sub-topic areas contained within an overarching domain.

ToolBus' abilities in addition to the implications of its design are:

- *Scalable*: Because there is no centralization of data sources, analysis tools, or even within the data models for information within a domain, ToolBus possesses no "brittleness" that would cause the addition of new tools or plug-ins to cause increasing time to develop, integrate, or debug. As the amount of information researchers examine increases, so too will client network bandwidth, though in a linear fashion. Only memory needs may grow in a greater than linear fashion due to the increase in possible groupings of information that would be possible.

- *Platform Independence*: All primary technologies upon which ToolBus is built are both hardware and operating system independent. Java run-time environments are available for an increasing number of different platforms allowing the ToolBus code and associated plug-ins to run nearly everywhere. We have developed on Intel/MSWindows, Intel/Linux, and Mac OS X to help ensure independence. By restricting ourselves to only primitive Java data types and Strings, Web-services can be contacted and utilized regardless of their host platform or implementation language. (This decision helps ensure that no advantage will be taken, as in passing references to more complex Java objects, that would violate platform independence and make any Web-services developed by us dependent upon the use of ToolBus. Thus, others can use our Web-services). XML documents, while verbose, provide a common format in which data can be understood and allow for easier examination by a human readership for debugging purposes if the need should arise. To enable ToolBus to have more information about Web-service operations, we utilize a specialized no-parameter operation that returns more detailed information about parameters and return types of other operations than can be found in the WSDL document for the service. We have been careful, however, to make use of this information only when available. Thus, such ToolBus friendly Web-services can work more seamlessly by performing parameter content type checking before bindings are attempted.
- *Domain Independence*: Although PathPort is originally being created within the life sciences domain, the need to later incorporate and integrate data from other disciplines was considered in the design and implementation. Although advantageous to create a discipline-specific system at times, the probability of a resulting brittle, non-scalable system as additional data types become needed would be greatly increased. The need to integrate other data types has already occurred. We now have plans to incorporate geographical information system (GIS) data into PathPort, and domain independence will make it plausible.
- *Rapid Visualization Development*: The development of new visualization applications can be time-consuming. ToolBus plug-ins can take advantage of numerous common elements to greatly reduce this development time. The tools managed by ToolBus provide varied means to provide results as XML documents to plug-ins, thus alleviating developers from the drudgery of supplying these input methods. Likewise the plug-in API of base classes provides support for view management, printing, user documentation/manual pop-ups, loading and saving of plug-in states as part of the ToolBus configuration, and presentation preferences. This allows developers to focus more of their efforts on visualization rather than recreating the common needs of all plug-ins.
- *Information Management*: As a client-side system, ToolBus gives users greater control over what information and analysis resources they use. Furthermore, since analysis tools will typically be accessed as Web-services, users or their organization—in the form of a local system administrator—do not need to maintain a suite of local analysis programs. Such maintenance tends to require either more knowledge on the part of the user, or runs the risk of being under-supported services, as is often the case in an academic

environment. An additional advantage, particular for small organizations, is ToolBus' support for lazy, rather than eager, data acquisition. Within the life sciences, the quantity of available data within databases such as GenBank [25] and SwissProt [26] is large. To improve the degree of integration with their tools, some organizations will reproduce an entire database locally. ToolBus alleviates the extra expense that duplicating such large quantities of data can incur by making these data available for easy download via Web-services.

- *Collaboration*: Users who invest large amounts of time finding, accumulating, analyzing, and mining information to solve problems should be able to save that work allowing them to restart where they left off and to share it with their collaborators anywhere in the world. ToolBus allows users to save and load their complete work-space state. Thus when a user loads a saved configuration, they get the same tools, models, groups, and views (including
the same window placement) as when they saved it, greatly aiding user reintegration into a saved task.
- *Grid, Web-portal, Existing Application Complimentary:* ToolBus is not meant to replace Grids. Instead, Web-services are points of entry into Grids (what has become known as Grid-services) [27]. The purpose of Web-services is to provide platform independence. If Grids are an effective way to support some Web-services, then they should be used like any technology. Web-services are also an excellent means to make platform-specific applications become platform-independent. In the case of existing executable programs, these can be wrapped as Web-services that convert text output to an appropriate XML document format. Such a process typically takes three to five days on average. The pages of Web portals can be wrapped in much the same way, though they are more prone to format changes in practice. Some Web-portals are beginning to wrap their various services as Web-services, e.g., EMBL [28].
- *Information Grouping, Comparison, and Combination*: The ability to create related groups of information and to interactively compare them allows the development of independent plug-ins, which in turn allows those plug-ins to be non-brittle by supporting the creation of more but smaller plug-ins. Comparisons are made by the interactive creation of Venn diagrams with users selecting those members to show or hide. Unfortunately, some domains can require the creation of groups with large memberships. To assist the user, we are currently working on a method to automatically discover interesting groups and to suggest their creation to the user. It is worthwhile to point out that plug-ins themselves might lead to the recognition of interesting results that can in turn lead to the instantiation of a new *Model*. ToolBus supports this type of feedback loop by allowing the *Model* to pass data to the ToolBus mediator class in exactly the same way that the *ToolManager* does.
- *Reusable Tools*: The data access and analysis tools within ToolBus can utilize one another if they are known. For example, an interactive query tool can be built on top of a stateless RPC style database access Web-service, giving the user a more user-friendly way of accessing a database. Likewise, a single analysis tool might be used by multiple plug-ins. This reduces overall development and subsequent maintenance time and costs for more complex and interactive tools and plug-ins.

3. Biochemical Modeling and Simulation

Genome projects have resulted in the completion of nearly 90 publicly available genomes by the end of 2002, with more than 120 others in progress. Genomes have been widely referred to as "parts lists" or "periodic tables" for living organisms. As such, they are static with respect to function, much like genetics can be considered constant with respect to the functional dynamics of living organisms (i.e., the "parts list" basically is stable during any given organism's lifetime). RNA and protein expression and the chemical (metabolite) composition of a living organism at a given moment in time under a specified environment provide information about the "state" of the organism. This is more closely related to the organism's biochemistry (the genome being its genetics). Although in flux, the biochemistry of living organisms can often be considered in quasi steady-state [29], [30].

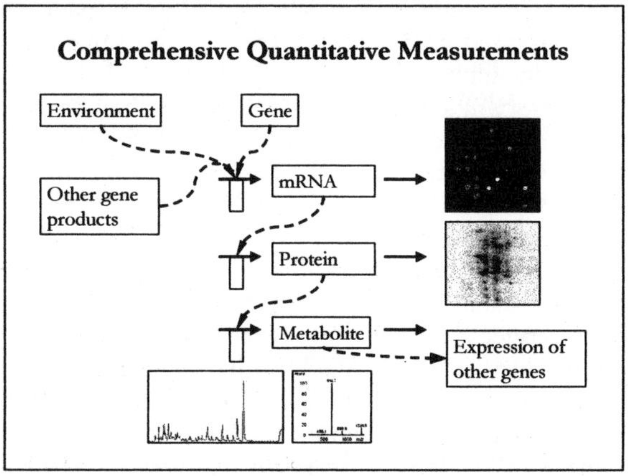

Figure 5. Comprehensive quantitative measurements can be made through various "Systems Biology Research Platforms" that analyze and measure DNA, RNA, proteins and metabolites. The resulting data can be integrated, managed, and analyzed via PathPort and associated tools.

Biochemistry must meet genetics to unravel the function of the genome in response to environmental variations, as information in the form of kinetic regulation is passed from the genome down to proteins and metabolites (as stated in the infamous "central dogma" of molecular biology) and visa-versa (Figure 5). Many proteins and protein complexes are regulators of gene expression. Certain metabolites are also modifiers of protein function, and modulators of gene expression (repressors or inducers). These various levels of molecular function are tightly bonded in a circular structure where neither of them are causes. Rather they all are effects, and the causes come only from the environment and pathogens. We do not yet have good methods with which to discover and assemble these connections and create predicative models—the major aim of biological sciences in the 21st century.

To increase our understanding of how living organisms function, an approach that joins experiments with computer modeling and theory must be realized. Fortunately, an existing, rich body of theoretical biochemistry is available on which to draw, such as stoichiometric analysis, thermodynamics, non-linear dynamics, statistical mechanics, metabolic control analysis, and control theory. A collection of mathematical tools is available, some older, such as systems of ordinary differential equations (ODEs), and some newer, such as Bayesian networks.

One approach, being pursued by Mendes and collaborators, is to take the "state data" (quantitative measurements of mRNA, proteins, and metabolites under specified and controlled conditions) and apply mathematical methods based on ODEs to infer networks [31], [32], [33]. The resulting network structures can then be studied mathematically. Referred to as reverse engineering, this is akin to what engineers do when they need to understand (usually to copy) an existing device for which they do not have specifications. The approach, by necessity, is a "top-down" strategy, where one starts from analyses of the *whole* system and thus uncovers some of its internal organization and complexity. Current areas of focus are inference of gene expression networks from mRNA profiling experiments and metabolic pathways from proteomics and metabolomics data. Laubenbacher and collaborators, working on the same problem, are using different mathematical methods—discrete mathematics and methods from symbolic computation. In the 1960s and 1970s, Kauffman proposed random Boolean networks as a discrete model for gene regulatory networks and other types of biochemical networks [34]. An intermediary between Boolean networks and continuous models such as ODEs is given by polynomial models over finite number systems. They allow the use of a well-developed algorithmic machinery for such problems as reverse-engineering of networks, reverse-engineering of dynamics, and simulation of systems.

A complementary and more established approach in mathematical modeling of cell regulatory systems, is sometimes thought of as "bottom-up". In this approach, data from purified molecules (mostly proteins) is gathered from the scientific literature and then used to generate testable hypotheses in the form of mathematical representations of biochemical networks [35], [36], [37]. These mathematical models are used in simulations to make predictions about the behavior of the system, which can then be put to the "gold standard" test of experimentation. The relationship(s) between the experimental results and the predictions is then fed back into the loop, helping to refine theory in an iterative process. The vision is to design and test hypotheses *in silico* using software (for example, *Gepasi*, available from Mendes at www.gepasi.org) that simulates the fundamental rules of biochemistry, i.e., chemical kinetics [38], [39]. Eventually, mathematical comprehension of individual reactions is built into a complete mechanistic understanding of the process, or what we originally understand by function [40].

Despite the large amount of available data, most biological systems of interest will be mathematically under-determined by the available state data. The presence of noise in most experimental data compounds this problem. As a result, no mathematical method alone can identify large biological networks and their dynamics correctly from experimental data alone, at least not in the near future. It must be aided by an understanding of the fundamental design principles of biological networks. In particular, we must understand how robustness and reliability features are integrated in such networks. Such an understanding can only come from a near comprehensive understanding of a select collection of model organisms. One example of work in this direction comes from Davidson *et al.* [41]. A complete picture of the wiring diagram of the gene regulatory network that directs embryonic development in the sea urchin is now available. This represents a crucial step toward a complete dynamic model of this network.

While existing mathematical methods for the understanding of biological networks have been quite successful in elucidating many observed phenomena, the challenges posed by computational biology call for the creation of whole new mathematical areas and the expansion of many existing ones. In particular, a comprehensive mathematical foundation for complex systems theory needs to be developed. The new mathematical area of data mining, once matured, will play a central role in uncovering patterns and relationships in existing data. The impact of these developments on mathematics will be comparable to that of the space program in the 1960s.

4. The Mathematics of Biological Networks

Mathematics provides a language that is particularly useful for solving problems across different domains and disciplines. Some of the mathematics needed for understanding how living organisms function is derived from graph theory, which has its origin in a paper by Euler [42]. Graph theory is ideally suited as a mathematical language for reasoning about connections between entities, such as networks of all types. The mathematical theory of networks, as it exists at present, is largely a branch of graph theory. In the 1950s, graph theory advanced through a series of eight seminal papers by Rényi and Erdos, introducing the new subject of random graph theory [43]. Their view supported random formation and essentially static networks.

One of the first instances of a large-scale study of the connectivity properties of real world networks was by Milgram. It gave rise to the famous "six-degrees-of-separation" metaphor [44]. Non-random connectivity in networks can result in what are referred to as "small world effects" because you can traverse large distances through few connections, involving nodes that have special properties such as high degrees of connection. Watts and Strogatz added to non-random connections by showing that real world networks might also be clustered [45]. For example, the likelihood that two of your colleagues know each other is high—suggesting that the social network formed by your friends is clustered and connected non-randomly. The current airport hub system offers another example. Long-range links between distant nodes in the system offer connections between distant hubs. Because of hubs you can fly between Los Angeles and New York on a single flight. An early example of how hubs relate to biology is the network of 302 neurons of a nematode (*Caenorhabditis elegans)*, which are connected in this way. The existence of these hubs or connectors in social networks is demonstrated through the recent observations of Gladwell [46]. In summary, biologically relevant networks do not seem to fit the random Rényi-Erdos networks because those networks do not support the existence of such connectors.

Like the airport hub system, many complex networks including many biologically relevant networks seem to obey power laws instead. This means that a few hubs get most of the action. Protein networks seem to fall into this category. These networks are referred to as scale-free. One process that creates these networks is growth of the network through preferential attachment to highly connected nodes or hubs [47]. The Internet is another example of such a scale-free network. Bianconi and Barabasi [48] have characterized these networks as "winner takes all," "fit get rich," and "first mover advantage," modeled after the dynamical processes of quantum gas phenomena.

The study of scale-free networks with mathematical tools that has emerged recently is an important beginning in the process of creating a new theory of biological networks [49]. It addresses the structure of the "wiring diagrams" that underlie complex systems such as biological networks. The next step is a comprehensive theory that allows the study of dynamics unfolding on this "backcloth". So, what questions would we like to ask of these biological networks as we infer them? Some examples pertain to their structure and how it varies, the relationship between these structures and the dynamics of living organisms, how to reconstruct them from partial data, the discovery of dynamic interaction patterns, and an understanding of how global dynamics are determined by local interactions.

5. Conclusions

Modern biology now stands at the threshold of understanding living organisms via a systems approach. The success of molecular biology and the numerous sequenced genomes have resulted in enormous amounts of available data, ripe to be analyzed via novel simulation, modeling, and mathematical tools. A comprehensive understanding of the various data types in a systems view now drives computational infrastructure development. The need to acquire, manage, analyze, and display large amounts of data as well as make useful predictions about the properties and/or behavior of living organisms in response to environmental and other perturbations will require scalable and flexible computational infrastructure. Infectious disease research now requires the development of such computational infrastructure. PathPort, short for Pathogen Portal, offers a web-based portal approach to the research community that will enable vertical and horizontal data integration. PathPort, a collaborative software development effort designed to develop, test, and implement a robust infrastructure for data management, integration, visualization, and sharing, is being built as a scalable portal providing data management and analysis for key pathosystems, especially relating to molecular data (e.g., genomics, metabolomics, transcriptomics, and proteomics). This approach may contribute to the decompartmentalization of scientists and funding agencies, which currently hinders the utilization of infectious disease data. The study of networks, both via novel simulation and mathematical modeling approaches, will also catalyze our understanding of biological systems. Virginia Bioinformatics Institute looks forward to further collaboration to advance each of these research areas.

Footnotes

[1]Dr. Bruno Sobral is corresponding author of this manuscript.

Acknowledgements

We thank the leadership of the Commonwealth of Virginia and Virginia Polytechnic and State University for the opportunity to create and develop the Virginia Bioinformatics Institute. We are grateful for funding from the US Department of Defense (DAAD 13-02-C-0018) and the Commonwealth Technology Research Fund (Virginia, 02-0843-10) to Dr. Bruno Sobral, and to the National Science Foundation (DBI-0109732, BES-0120306, IBN-0118612) for funding to Dr. Pedro Mendes and collaborators.

Special thanks to Drs. Raju Lathigra and Yongqun He for providing leadership in data acquisition regarding pathogens of interest to the PathPort project and Eric Nordberg for significant contributions to the project. Many thanks to all members of the PathPort team for their efforts. Finally, we thank Dr. Neysa Call and Tiffany Trent for edits and reviews.

References

[1] S. Kauffman, Metabolic Stability and Epigenesis in Randomly Constructed Genetic Nets, *J. Theoret. Biol.* **22** (1969) 437-467.

[2] National Institute of Health, Bioinformatics at the NIH, Available at http://www.bisti.nih.gov, Accessed January 14, 2003.

[3] The Economist, The Race to Computerise Biology, December 12, 2002, Available at http://www.economist.com/science/tq/displayStory.cfm?story_id=1476685, Accessed January 14, 2003.

[4] I. Adelman, Computing with DNA, *Scientific American* **August** (1998) 54-61.

[5] J. Holland, Adaptation in Natural and Artificial Systems, University of Michigan Press, Ann Arbor, MI, 1975.

[6] M. Mitchell, An Introduction to Genetic Algorithms, MIT Press, Boston, MA, 1995.

[7] IBM Research, Blue Gene Project, Available at http://www.research.ibm.com/bluegene/, Accessed January 9, 2003.
[8] N. Wiener, Cybernetics: or Control and Communication in the Animal and the Machine, MIT Press, Cambridge, MA, 1961.
[9] Genome News Network, A Guide to Sequenced Genomes, Available at http://gnn.tigr.org/sequenced_genomes/genome_guide_p1.shtml, Accessed January 9, 2003.
[10] J. Heinrich, Bioterrorism Coordination and Preparedness (GAO-02-129T), Testimony by Director, Healthcare and Public Health Issues before Subcommittee on Government Efficiency, Financial Management, and Intergovernmental Relations, Committee on Government Reform, US House of Representatives, October 5, 2001.
[11] Virginia Bioinformatics Institute, The Pathogen Portal Web Project, Available at http://www.vbi.vt.edu/~pathport/, Accessed January 9, 2003.
[12] A. Grimshaw *et al.*, In: F. Berman *et al.*(eds.), Grid Computing: Making the Global Infrastructure a Reality, Wiley, 2003.
[13] Legion, Worldwide Virtual Computer, Available at http://legion.virginia.com/, Accessed January 14, 2003.
[14] I. Foster and C. Kesselman, The Globus Project: A Status Report, *Proc. Heterogenous Computing Workshop.* IEEE Press, 1998, 4-18.
[15] Biomed Grid Portal, BioInformatics Institute 2002, Available at http://bmg.bii.a-star.edu.sg/, Accessed January 9, 2003.
[16] BioASP, Available at http://www.bioasp.nl, Accessed January 9, 2003.
[17] Technical Computing Portal, Ohio Supercomputer Center, Bioinformatics Resources, Available at http://www.osc.edu/research/bioinformatics/portal/software.shtml, Accessed January 9, 2003.
[18] Zerosum Bioinformatics Web Portal, Available at http://www.zerosum.com/ZerosumPortal.htm, Accessed January 10, 2003.
[19] W. Venables and B. Ripley, Modern Applied Statistics with S-PLUS, Springer Verlag, 1999.
[20] International Business Machines Corporation, IBM Websphere Portal Server Product Architecture V2.1, Available at http://www-3.ibm.com/software/webservers/portal/pdf/WPSWhitePaperV2-1.pdf, Accessed January 9, 2003.
[21] Oracle Corporation, Available at http://portalstudio.oracle.com/pls/ops/docs/FOLDER/COMMUNITY/OTN_CONTENT/MAINPAGE/OWCOP, Accessed December 15, 2002.
[22] MolMine, Jexpress Pro, Available at http://www.molmine.com/frameset/frm_jexpress.htm, Accessed January 10, 2003.
[23] W3C, W3C Technical Reports and Publications, Available at http://www.w3.org/TR/#Recommendations, Accessed January 10, 2003.
[24] B. Boehm, A Spiral Model of Software Development and Enhancement, IEEE Computer, 1998.
[25] NCBI, GenBank Overview, Available at http://www.ncbi.nlm.nih.gov/Genbank/GenbankOverview.html, Accessed January 10, 2003.
[26] EMBL-EBI, SWISS-PROT, Available at http://www.ebi.ac.uk/swissprot/index.html, Accessed January 10, 2003.
[27] I. Foster, C. Kesselman, J. Nick, S. Tuecke, Grid Services for Distributed System Integration, IEEE Computer, 2002, pp. 37-46.
[28] European Molecular Biology Laboratory, Research in Molecular Biology, Available at http://www1.embl-heidelberg.de/, Accessed January 14, 2003.
[29] L. von Bertalanffy, Theoretische Biologie (Theoretical Biology), Berlin, 1942.
[30] I. Prigogine, Introduction to Thermodynamics of Irreversible Processes, Wiley, New York, 1961.
[31] A. de la Fuente, P. Brazhnik, and P. Mendes, A quantitative method for reverse engineering gene networks from microarray experiments using regulatory strengths, *Proceedings of the 2nd International Conference on Systems Biology*, California Institute of Technology, Pasadena, CA, 2001.
[32] A. de la Fuente and P. Mendes. Quantifying gene networks with regulatory strengths, *Molecular Biology Reports* **29**, 2002, 73-77.
[33] A. de la Fuente, P. Brazhnik, P. Mendes, Linking the genes: inferring quantitative gene networks from microarray data, *Trends in Genetics* **18**, (2002), 395-398.
[34] S. Kauffman, Gene regulation networks: a theory for their global structure and behavior, *Current Topics in Dev. Bio.* **6** (1971) 145.
[35] D. Garfinkel, Computer modeling of metabolic pathways, *Trends Biochem. Sci.* **6** (1981) 69-71.
[36] J. Reich and E. Selkov, Energy Metabolism of the Cell: a Theoretical Treatise, Academic Press, London, 1981.
[37] J. Hofmeyr, Steady-state Modeling of Metabolic Pathways: a Guide for the Prospective Simulator, *Comput. Appl. Biosci.* **2** (1986) 5-11.

[38] P. Mendes, GEPASI: a Software Package for Modeling the Dynamics, Steady States, and Control of Biochemical and Other Systems, *Comput. Appl. Biosci.* **9** (1993) 563-571.
[39] P. Mendes, Biochemistry by Numbers: Simulation of Biochemical Pathways with Gepasi 3, *Trends Biochem. Sci.* **22** (1997) 361-363.
[40] P. Brazhnik, A. de la Fuente, and P. Mendes, Gene Networks: How to Put the Function in Genomics, *Trends Biotechnol.* **20** (2002) 467-472.
[41] E. Davidson, *et al.*, A Genomic Regulatory Network for Development, *Science* **295** (2002) 1669-1678.
[42] L. Euler, The Solution of a Problem Relating to the Geometry of Position, 1736.
[43] P. Erdos and A. Renyi, On Random Graphs: I, *Math. Debrecen* **6** (1959) 290-297.
[44] S. Milgram, The Small World Problem, *Psychology Today*, May 1967, 60-67.
[45] D. Watts and S. Strogatz, Collective Dynamics of 'Small-world' Networks, *Nature* **393** (1998), 440-442.
[46] M. Gladwell, The Tipping Point: How Little Things Can Make a Big Difference, Little, Brown, & Company, 2000.
[47] A. Barabasi and R. Albert, Emergence of Scaling in Random Networks, *Science* **286** (1999) 509-512.
[48] G. Bianconi and A. Barabasi, Bose-Einstein Condensation in Complex Networks, *Physical Review Letters* **86** (2001), 5632-5635.
[49] A Barabasi, Linked: The New Science of Networks, Perseus Publishing, Cambridge, MA, 2002.

Toxicogenomics and Proteomics
J.J. Valdes and J.W. Sekowski (Eds.)
IOS Press, 2004

Interpretation of Global Gene Expression Data through Gene Ontologies

Jacques RETIEF, Joe MORRIS, Jill CHENG, Tarif AWAD, Rusty THOMAS* and John HOGENESCH*
Affymetrix, Inc. 3380 Central Expressway, Santa Clara, CA 95051, USA
**Genomics Institute of the Novartis Foundation, 10675 John Jay Hopkins Drive, San Diego, CA 92121, USA*

Abstract. The new field of toxicogenomics, with the aid of genome-wide arrays, Promises to deliver a complete view of all the molecular processes in the cell. Such a complete view should enable us to simultaneously monitor the on-target effect of a drug as well as itstoxicological side effects and other perturbations of the biological system. To deliver on this promise we need a robust experimental design capable of detecting small, but statistically significant, changes in the gene expression level combined with a biological understanding of the gene functions. We will discuss the incorporation of Gene Ontology, (GO) classifications with gene expression data to understand the mechanism of action of a well-studied drug.

1. Background

Large databases of drug profiles are increasingly being used to predict unintended effects, such as off-target effects and toxicological side effects, of new drugs. In this application the expression patterns of a new drug are compared to the expression profiles of a series of drugs from known drug classes with known side effects. To predict possible side effects it is important that the database must be large enough to cover all known drug classes and the micro-arrays provide an extensive representation of all known genes. The use of function annotations, such as GO ontologies, provides an additional avenue to extract more useful information from a drug expression profile database. This approach offers a number of benefits:

a. Previously unknown side effects may be predicted based on function. In other words, we may begin to understand the mechanism of a side effect that was not represented by any of the drugs already in the database. A logical extension of this approach is that we can make better predictions with smaller databases.
b. Annotations provide insight into mechanism of action. Once we understand or postulate a mechanism of action we can focus on appropriate biological models to evaluate it further. A detailed knowledge of mechanism of action is also highly desirable for regulatory approval.
c. Knowledge of biological functions is transferable between models. Without any knowledge of the function of expression patterns, it is very difficult to relate the significance of a particular expression pattern change to a different animal model, or even to a different animal strain.
d. Annotations simplify interpretation. Expression patterns are complex. Annotations provide a unifying framework to group expression patterns in a way that is easy for a biologist to interpret.

2. Materials and Methods

The drug used in this experiment, 1,5-Isoquinolinediol (ISQ), is a specific and potent poly(ADP-ribose)polymerase (PARP) inhibitor[2].

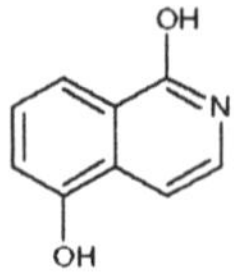

Figure 1. 1,5-Isoquinolinediol

The drug target, poly(ADP-ribose)polymerase (PARP) is an abundant nuclear protein which is activated by DNA breaks. Once activated, the enzyme catalyses the synthesis of poly(ADP-ribose) from NAD^+ on a number of proteins such as histones, ligases, topoisomerases and PARP itself [4-6]. For every molecule of NAD consumed, four free energy equivalents of ATP are required to restore NAD levels. It is possible that this excessive ATP consumption may trigger apoptosis by opening the mitochondrial permeability transition core [7]. The ability to inhibit PARP and hence apoptosis have stimulated interest in using apoptosis inhibitors as therapeutic agents in diverse diseases [2,8-10], such as:

a. Myocardial reperfusion injury, cardiomyopathy, toxic myocardial injury, and diabetic cardiovascular dysfunction.
b. Stroke and neurotrauma.
c. Reperfusion injury of the gut, eye, kidney, and skeletal muscle.
d. Arthritis.
e. Inflammatory bowel disease and inflammatory diseases of the central nervous system.
f. Multiple sclerosis.
g. Adjuvant therapeutics for the treatment of various forms of cancer.

The biological system is composed of Human HepG2 cells which were exposed for 4 hours at two concentrations, 1µM and 10µM of ISQ. The drug delivery vehicle, DMSO, was used as a control. Total RNA was isolated from each treatment condition; the samples were labeled in duplicate and hybridized to Affymetrix HG_U133A arrays.

The on-target effect of ISQ is to inhibit the up-regulation of PARP which is induced by DNA strand breakage. In this experiment the human HepG2 cells are not challenged by a DNA-breaking agent, so we do not expect to see the on-target effect of ISQ. Instead, we are focusing on off-target and toxicological effects.

The gene ontologies were derived from the GO database (http://www.geneontology.org/) [11]. Biological pathways were mapped with the Gene MicroArray Pathway Profiler, GenMAPP (http://www.genmapp.org/)[12].

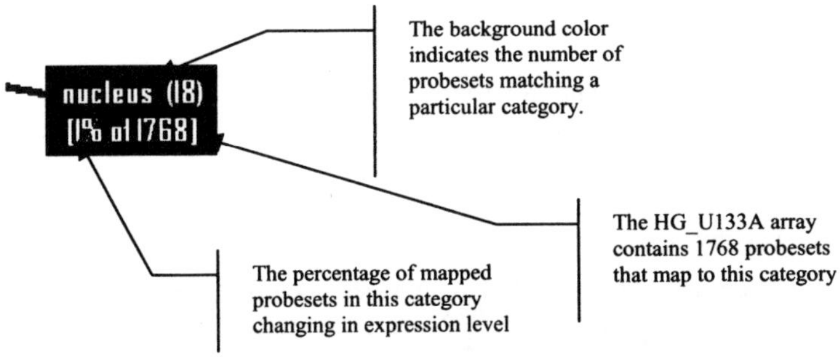

Figure 2: The GO figures are drawn from the "trunk" on the left, which represents the most general category, to progressively more specific categories, "leaves" on the right. The significance of the change in expression level is usually reflected in the length of the branches. For example, a very general category, close to the trunk will have many probesets mapped to the array. We expect a certain number of probes to match such a big, general category purely by chance. As we proceed to the leaves on the right, the categories become relatively smaller and more specific. It is much less likely that a group of probes will match such a small category by chance. This is reflected in the % value that tends to increase to the right.

The data analysis was carried out as follows: vehicle (DMSO)–only expression levels were compared to two concentrations of ISQ in duplicate. Increase and decrease calls were generated by Affymetrix Micro Array Suite V5.0. Lists of probesets were generated by looking for consistent increase or decrease calls between all DMSO and ISQ treated cells.

3. Results

There are two significant functional categories that are up-regulated: TRANSCRIPTION and NEGATIVE CONTROL OF CELL PROLIFERATION (Figure 3.). First, the transcription process, mostly involving Pol II is implicated. This is a general reaction common to eukaryotic transcription and may reflect a general mobilization of the genome.

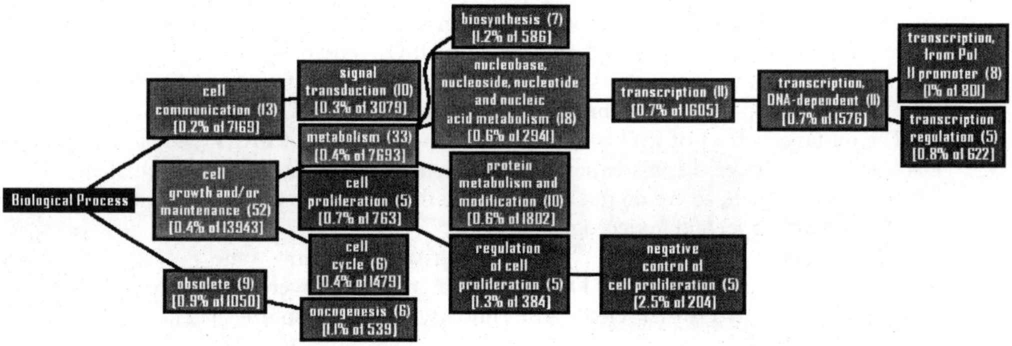

Figure 3. Ontologies of transcripts that increase in abundance.

The NEGATIVE CONTROL OF CELL PROLIFERATION is the most significant category that is increased in expression level (Figure 3.). It has been reported that when NB4 cells when treated simultaneously with all trans retinoic acid and ISQ failed to differentiate into neutrophils [13]. This effect has been attributed to the down regulation of PARP, but the off-target effect noted here may be a contributing factor.

Table 1. Details of the probesets up-regulated in the two most specific, or significant, categories.

GO: 8285 negative control of cell proliferation			
probeset	**function**	**gene**	**locus**
201218_at	C-terminal binding protein 2	CTBP2	10q26.13
202704_at	transducer of ERBB2, 1	TOB1	17q21
204159_at	cyclin-dependent kinase inhibitor 2C (p18, inhibits CDK4)	CDKN2C	1p32
210371_s_at	retinoblastoma binding protein 4	RBBP4	1p34.3
208654_s_at	CD164 antigen, sialomucin	CD164	6q21
GO:6357 transcription regulation from Pol II promoter (experimental evidence)			
probeset	**function**	**gene**	**locus**
209187_at	down-regulator of transcription 1, TBP-binding (negative cofactor 2)	DR1	1p22.1
221727_at	activated RNA polymerase II transcription cofactor 4	PC4	5p13.3
200047_s_at	YY1 transcription factor	YY1	14q
212761_at	transcription factor 7-like 2 (T-cell specific, HMG-box)	TCF7L2	10q25.3
202396_at	transcription elongation regulator 1 (CA150)	TCERG1	5q31
209377_s_at	thyroid hormone receptor interactor 7	TRIP7	6q15
201652_at	COP9 constitutive photomorphogenic homolog subunit 5 (Arabidopsis)	COPS5	8q12.3
202370_s_at	core-binding factor, beta subunit	CBFB	16q22.1

Mapping expression results to an ontology has the added benefit of filtering out some false positives. The filter threshold is set by the requirement that for every functional category at least 3 transcripts change in expression level. A false positive will be expected to map to a random category. That random category it will not appear in the figure if fewer than 3 transcripts map to it. This allows us to relax our statistical filters and increase sensitivity. The view presented by the GO ontologies depends on the statistical parameters used to select the list of genes that are changed. Figure 4 panel B represents a pathway built on a relaxed statistical stringency where 75% of the replicates show an increase in expression level, instead of 100% showing an increase when compared to the vehicle (Figure 4, panel A). The major pathway is the same.

The biological processes of metabolism and biosynthesis are down regulated, consistent with a general slowing, or shutting down of the cell metabolism.

The additional category, DEVELOPMENTAL PROCESSES that is shown to be down regulated in panel B, is substantiated by the up-regulation of the converse process, NEGATIVE CONTROL OF CELL PROLIFERATION (Figure 3.).

A

B

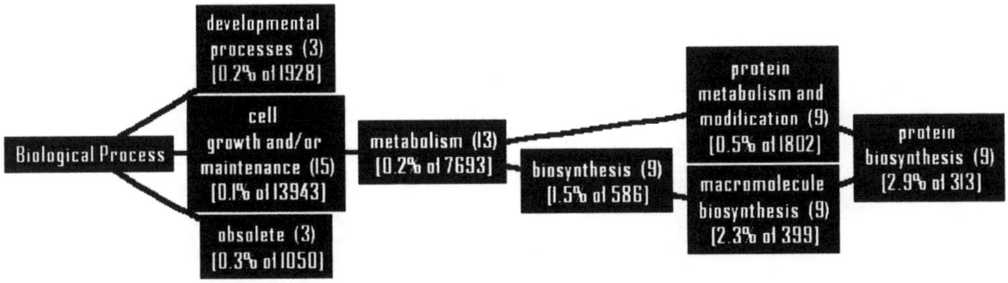

Figure 4. Ontologies of transcripts that decrease in abundance.
The view presented by the GO ontologies depends on the statistical parameters used to select the list of genes that are changed. Panel A. 100% of the replicates show an increase in expression level when compared to the vehicle Panel B represents a pathway built on a relaxed statistical stringency, where 75% of the replicates show an increase in expression. The major pathway is the same.

4. Discussion

Isoquinolinediol, ISQ is known to be a potent and highly specific poly(ADP-ribose) polymerase inhibitor. Under the experimental conditions used the PARP expression levels are not elevated, so the effect of ISQ is muted and our observations focused on off-target and toxicological effects. Very few such effects have been noted previously and this study confirms those findings.

The ontologies proved to be a powerful tool to make the data readily interpretable by a biologist. Most of the biological processes found were consistent with processes we expect to find in cells under the conditions tested.

The finding of an increase in the negative control of cell proliferation is unusual. We did not see any differences in the expression levels of PARP-1, so this effect may be due to a number of factors:

a. Changes in the PARP-1 enzyme activity due to:
   - Post-translational modification [14]
   - Auto-poly(ADP)-ribosylation. [15]
   - Phosphorylation [16]
b. An off-target effect. ISQ itself may trigger the expression of a number of tumor suppressor genes. Further experiments will have to be carried out to distinguish between the effect of PARP and ISQ in the changes in expression levels of these genes.

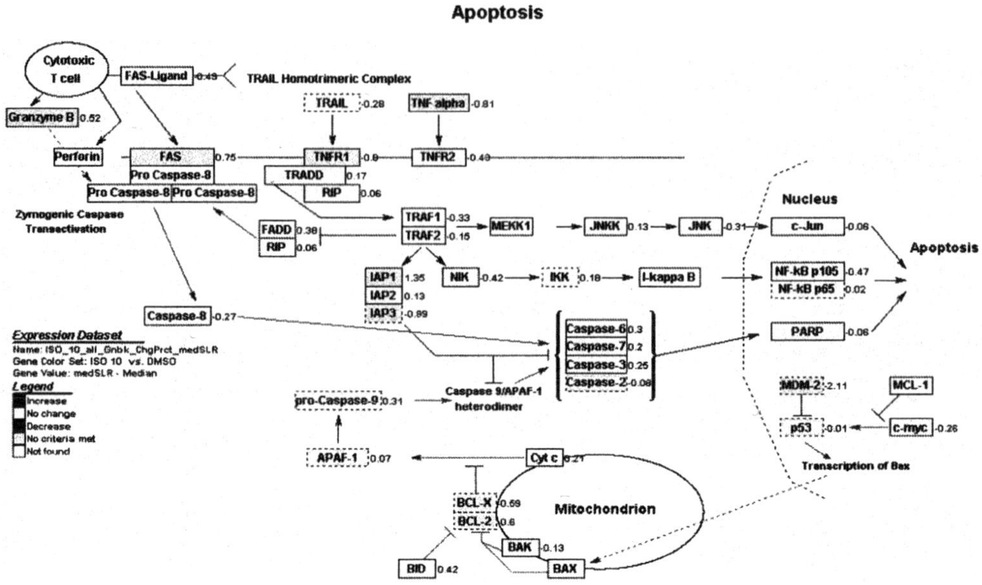

Figure 5. GenMAPP provides a detailed view of the apoptosis pathway. It confirms that, under the experimental conditions used, there are no DNA strand breaks and hence no triggering of the apoptosis pathway. Under these conditions it is unlikely that PARP-1 activity is modulated. Please note: this pathway predates the discovery of AIF [1] and the publication of the Caspase-independent apoptosis pathway [3].

In the absence of direct evidence apoptosis activity can be used as indirect evidence to distinguish between changes in the PARP activity and off-target effects of ISQ. Changes in PARP-1 activity is known to trigger apoptosis [7]. However, we find no changes in the expression levels of any of the mapped apoptosis genes, so there is no indirect evidence for modulation of PARP-1 activity.

5. Conclusions

This study highlights the of global gene profiling in conjunction with annotations to generate hypotheses. The increase value in expression levels of tumor suppressor genes would have been hard to detect *a priori*, because the effect would be masked by changes in the stability of the genome produced by changes in PARP expression levels. Also, the additional information about apoptosis levels helped us to distinguish between ISQ and PARP-effects. Clearly, further experiments are needed to confirm and optimize the anti-tumor potential of this drug.

References

[1] S.W. Yu, H. Wang, M.F. Poitras, C. Coombs, W.J. Bowers, H.J. Federoff, G.G. Poirier, T.M. Dawson and V.L. Dawson Mediation of poly(ADP-ribose) polymerase-1-dependent cell death by apoptosis-inducing factor, Science 297 (2002) 259-263.

[2] L. Virag and C. Szabo The therapeutic potential of poly(ADP-Ribose) polymerase inhibitors, Pharmacol Rev 54 (2002) 375-429.

[3] S.P. Cregan, A. Fortin, J.G. MacLaurin, S.M. Callaghan, F. Cecconi, S.W. Yu, T.M. Dawson, V.L. Dawson, D.S. Park, G. Kroemer and R.S. Slack Apoptosis-inducing factor is involved in the regulation of caspase-independent neuronal cell death, J Cell Biol 158 (2002) 507-517.
[4] A. Chiarugi and M.A. Moskowitz Cell biology. PARP-1--a perpetrator of apoptotic cell death?, Science 297 (2002) 200-201.
[5] L. Tentori, I. Portarena and G. Graziani Potential clinical applications of poly(ADP-ribose) polymerase (PARP) inhibitors, Pharmacol Res 45 (2002) 73-85.
[6] A. Semionov, D. Cournoyer and T.Y. Chow Inhibition of poly(ADP-ribose)polymerase stimulates extrachromosomal homologous recombination in mouse Ltk-fibroblasts, Nucleic Acids Res 27 (1999) 4526-4531.
[7] F. Di Lisa, R. Menabo, M. Canton, M. Barile and P. Bernardi Opening of the mitochondrial permeability transition pore causes depletion of mitochondrial and cytosolic NAD+ and is a causative event in the death of myocytes in postischemic reperfusion of the heart, J Biol Chem 276 (2001) 2571-2575.
[8] J.C. Reed Apoptosis-based therapies, Nat Rev Drug Discov 1 (2002) 111-121.
[9] D.W. Nicholson From bench to clinic with apoptosis-based therapeutic agents, Nature 407 (2000) 810-816.
[10] J. Yuan and B.A. Yankner Apoptosis in the nervous system, Nature 407 (2000) 802-809.
[11] M. Ashburner, C.A. Ball, J.A. Blake, D. Botstein, H. Butler, J.M. Cherry, A.P. Davis, K. Dolinski, S.S. Dwight, J.T. Eppig, M.A. Harris, D.P. Hill, L. Issel-Tarver, A. Kasarskis, S. Lewis, J.C. Matese, J.E. Richardson, M. Ringwald, G.M. Rubin and G. Sherlock Gene ontology: tool for the unification of biology. The Gene Ontology Consortium, Nat Genet 25 (2000) 25-29.
[12] K.D. Dahlquist, N. Salomonis, K. Vranizan, S.C. Lawlor and B.R. Conklin GenMAPP, a new tool for viewing and analyzing microarray data on biological pathways, Nat Genet 31 (2002) 19-20.
[13] D.M. Berry, K. Williams and K.A. Meckling-Gill All trans retinoic acid induces apoptosis in acute promyelocytic NB4 cells when combined with isoquinolinediol, a poly(ADP-ribose) polymerase inhibitor, Leuk Res 24 (2000) 307-316.
[14] G. Doucet-Chabeaud, C. Godon, C. Brutesco, G. de Murcia and M. Kazmaier Ionising radiation induces the expression of PARP-1 and PARP-2 genes in Arabidopsis, Mol Genet Genomics 265 (2001) 954-963.
[15] M. Kawaichi, K. Ueda and O. Hayaishi Multiple autopoly(ADP-ribosyl)ation of rat liver poly(ADP-ribose) synthetase. Mode of modification and properties of automodified synthetase, J Biol Chem 256 (1981) 9483-9489.
[16] P.I. Bauer, K.G. Buki and E. Kun Selective augmentation of histone H1 phosphorylation sites by interaction of poly(ADP-ribose) polymerase and cdc2-kinase: comparison with protein kinase C, Int J Mol Med 8 (2001) 691-693.

Toxicogenomics and Proteomics
J.J. Valdes and J.W. Sekowski (Eds.)
IOS Press, 2004

Expanding the Information Window to Increase Proteomic Sensitivity and Selectivity

Paul SKIPP[1], Mateen FAROOQUI[2], Karen PICKARD[1], Yan LI[1],
Alan G. R. EVANS[2] and C. David O'CONNOR[1]
[1]*Centre for Proteomic Research and School of Biological Sciences, University of Southampton, Southampton SO16 7PX, UK*
[2]*Microelectronics Research Group, Department of Electronics & Computer Science, University of Southampton,*
Southampton SO16 1BJ, UK

Abstract. Proteomics is already invaluable for both applied studies, e.g. the discovery of novel biomarkers, and in fundamental investigations of cellular processes. However, there is a clear need to increase its sensitivity and selectivity, while retaining a high throughput of samples. We report here two general approaches: (i) systematic use of orthogonal mass datasets via methylation and (ii) enhanced chips for Desorption/Ionisation on Silicon (DIOS), which go some way to address these issues. In the first procedure, samples for peptide mass fingerprinting are split and one half is methylated at COOH groups. The masses of the native and methylated peptides are then measured and used to search sequence databases in conjunction with a modified mass fingerprinting algorithm. Such orthogonal mass datasets efficiently identify proteins even when very limited peptide mass data is available. The second approach uses a novel DIOS chip fabricated in single crystal silicon to circumvent the problem of chemical noise inherent in matrix-assisted laser desorption ionisation mass spectrometry (MALDI-MS). The chips have a hitherto unreported columnar structure with an extremely high aspect ratio that can be reproducibly fabricated in localised areas by masking. The columnar structure has been realised without the use of pore initiation techniques, which require high definition lithography, and is quite different from porous silicon obtained by electrochemical etching techniques. The use of both approaches in several biological settings, including the analysis of the pathogen *Mycobacterium tuberculosis*, is described.

1. Introduction

Diagnostic patterns of protein expression – proteomic signatures – show considerable potential in both fundamental and applied fields of biomedical science. In terms of basic science, such signatures can be used to define particular physiological states in cells or tissues and may also implicate specific proteins in key cellular processes. Similarly, they can be of use in applied studies for the early detection of diseases or their causative agents, e.g. toxins, bacteria and viruses.

Generally, measurement of the level of any one biomarker protein is not highly predictive of a physiological process or the presence of a specific type of harmful agent. However, recent studies suggest that analysis of the levels of several biomarkers in combination can be highly discriminatory (see e.g. [1, 2]). It is therefore crucial to uncover as many biomarkers as possible to derive reliable proteomic signatures. For these and other reasons, there is currently much effort to improve proteomic technologies so that they can sample as much of the expressed proteome as possible.

Proteome sampling presents formidable challenges. For example, proteins have extremely diverse physico-chemical properties, which mean that it is difficult to employ a single approach to separate and detect all expressed polypeptides. Additionally, the dynamic range of protein expression is vast, e.g. from ~1 to ~10^9 molecules per ml for serum [3]. Even so, many potential biomarker proteins are present at relatively low levels. It is therefore difficult to monitor such proteins without time-consuming enrichment procedures.

Low abundance proteins are also frequently difficult to identify by high throughput methods such as peptide mass fingerprinting. In this procedure, peptides from the protein of interest are produced by treatment with a site-specific endoprotease such as trypsin and their masses measured, typically by matrix-assisted laser desorption/ionisation-mass spectrometry (MALDI-MS). The experimental peptide mass data is then used to search for matching proteins using a database consisting of a 'virtual digest' of all known sequences [4-6]. Low abundance proteins present problems with this technique because they only yield a limited subset of peptides after protease treatment and extraction. This thwarts attempts to derive unambiguous matches with candidate proteins in subsequent database searches. It is therefore frequently necessary to derive sequence tags or even *de novo* amino acid sequence data, both of which are relatively slow and low throughput procedures [7, 8].

In this paper, we describe two approaches to address these problems. In the first, we show that the systematic use of an orthogonal mass dataset, obtained after methylation of one-half of a peptide sample, greatly improves the discriminatory power of peptide mass fingerprinting, even where limited numbers of individual peptides are recovered. The procedure also has clear advantages in the identification of small proteins, where direct protein identification is also impaired by the restricted number of endoproteolytic peptides.

The second approach focuses on a limitation of MALDI-MS technology in small molecule analysis, which is a consequence of the need to add an organic acid matrix to samples. This creates molecular noise at the lower end of the mass spectrum, thereby narrowing the 'information window'. In 1999, Wei *et al.* reported Desorption/Ionisation on Silicon (DIOS), which circumvents this problem by employing electrochemically-etched porous silicon to desorb samples without matrix [9]. The technique allows the measurement of molecules with *m/z* values of <700, simplifies sample preparation and has a higher salt tolerance than MALDI-MS. However, the performance characteristics of such silicon chips are somewhat variable. It is already evident that reproducible and controlled fabrication of DIOS surfaces will be essential for optimisation and routine use of this technology. In the second approach, therefore, we developed enhanced DIOS chips with highly regular columnar structures of predetermined length. The highly reproducible fabrication of these structures in localised areas defined by oxide masking obviates the need for pore initiation techniques and consequent high definition photolithography. The usefulness of both approaches is illustrated by examples of biological relevance, including the identification of proteins of the bacterial pathogen *Mycobacterium tuberculosis* that are differentially expressed on host infection.

2. Materials and Methods

Preparation of protein extracts and two-dimensional gel electrophoresis

A detailed description of the method used to recover *Mycobacterium tuberculosis* from mouse models of acute and chronic infection will be described elsewhere. Preparation of whole cell extracts of *M. tuberculosis* and fractionation on 2-dimensional gels were carried out essentially as previously described [10-13]. Gels were silver-stained [14] or, in some cases, stained with SYPRO Ruby. Image analysis was carried out as described [11]

Spot recovery, in situ digestion of proteins and peptide extraction

Protein spots of interest, identified by image analysis using Phoretix 2-D or PDQUEST software, were excised from gels by hand or by using a BioRad robotic spot-cutter. Protein alkylation, *in situ* digestion with trypsin and peptide recovery were carried out manually as previously described [11] or with a MassPrep workstation (Micromass) prior to sample clean-up on a Waters 2700 workstation.

Peptide methylation

100 µl of acetyl chloride were added to 1 ml of -20°C methanol and allowed to warm to room temperature. 5 µl of the solution were then added to lyophilised tryptic peptides and, after 1 hour, the sample was re-lyophilised. Following resuspension in 1 % formic acid, the sample was desalted on a C18 microcolumn (a 50:50 mixture of Poros R2 and R3 resin) as previously described [11, 15]. The washed sample was eluted into nanospray capillaries.

MALDI- and DIOS-mass spectrometry

Standard procedures and matrices were used for MALDI-MS. Samples were generally dissolved in 0.1% trifluoroacetic acid prior to application of 1 µl to a sample plate and mixing with 1 µl of matrix. Preparation of the enhanced DIOS chips will be described elsewhere. For use with a laser desorption/ionisation mass spectrometer, the chips were mounted on standard MALDI sample plates that had been machined to ensure that the DIOS surface was flush with the rest of the plate. DIOS and MALDI samples were analysed on a M@LDI mass spectrometer (Micromass) using the following settings: pulse voltage: 2500V; detector: 2000V; laser: 50 – 65%.

Nanoelectrospray mass spectrometry

Desalted peptides were collected into borosilicate nanospray capillaries by centrifugation prior to nanoelectrospray MS using an LCT orthogonal acceleration-TOF instrument (Micromass), with the following settings: capillary voltage, 1200V; cone voltage, 46-52V; desolvation gas, 100-200 L/h; source temperature, 80°C.

Database searching

Experimentally-derived masses of tryptic peptides were used to search an on-line non-redundant National Center for Biotechnology Information (NCBI) database for matching proteins using an in-house version of MS-Fit in the Protein Prospector software suite (also available at: http://prospector.ucsf.edu/). For methylated peptides, MS-Fit was modified so that the masses of Asp and Glu residues were both increased by 1C and 3H atoms.

Other procedures

Scanning electron microscopy of enhanced DIOS chips used standard procedures. The chips were also photographed under UV light using a VersaDoc image station (BioRad).

3. Results

Systematic use of orthogonal mass datasets for increased accuracy of protein identification

The many genome sequencing projects that have been completed have generated a correspondingly vast number of potential protein sequences that are available for searching in databases such as the NCBI Protein database, SWISS-PROT and TrEMBL. An important consequence of this increase in information is that protein identification approaches such as peptide mass fingerprinting are now much more likely to yield false positives than previously.

To increase confidence in protein identification, we reasoned that the systematic use of a complementary mass dataset, in combination with the traditional peptide mass fingerprinting approach, would significantly reduce the problem. A simple way to generate such a dataset would be to methylate the COOH groups in peptides prior to searching the databases with a suitably modified mass fingerprinting algorithm. If the 'hit lists' obtained with the normal and methylated peptides both uncovered the same protein, then it would be exceedingly unlikely that this was a false positive. For example, if the probability of obtaining a false positive on either list is 1×10^{-3} then the probability of obtaining the same false positive on both lists is the product of the individual probabilities (in this case, 10^{-6}).

To validate the idea, we first tested known proteins. Table 1 shows that mass fingerprinting of normal and methylated peptides of bovine carbonic anhydrase correctly identified it in both cases. Significantly, no other proteins in the first 100 candidate proteins were found in both lists. It should be noted, however, that analysis of the non-methylated sample ranked the bovine and sheep forms of carbonic anhydrase in joint first place. (The latter had a greater number of matching peptides but also had a lower MOWSE score). Only when the methylated sample results were available was it possible to discriminate between them. Similar results were obtained for other characterised proteins.

Table 1: Validation of orthogonal mass dataset approach – results obtained after search with MS-Fit or modified MS-Fit, using normal and methylated peptide mass data for bovine carbonic anhydrase

RANK	MOWSE Score	Number of masses matched (%)	Protein mass (Da)/pI	NCBI accession number	Protein name (species)
Normal peptides					
1	1.07×10^7	12/52 (23%)	29403.5/6.4	115453	carbonic anhydrase II (bovine)
2	8.69×10^5	13/52 (25%)	29514.7/6.4	115460	carbonic anhydrase II (sheep)
3	4.06×10^3	8/52 (15%)	88954.0/5.08	84897	(X51652) Bic-D protein (*Drosophila melanogaster*)
3	4.06×10^3	8/52 (15%)	88954.0/5.08	114985	(M31684) bicaudalD protein (*Drosophila melanogaster*)
4	1.94×10^3	6/52 (11%)	93085.5/5.63	520476	(U09129) Mkt1p (*Saccharomyces cerevisiae*)
Methylated peptides					
1	2.17×10^6	10/38 (26%)	29403.5/6.4	115453	carbonic anhydrase II (bovine)
2	1.01×10^5	8/38 (21%)	29514.7/6.4	115460	carbonic anhydrase II (sheep)
3	2.43×10^3	5/38 (13%)	45766.4/8.59	2493896	(X81973) cysteine synthase (*Arabidopsis thaliana*)
4	1.63×10^3	6/38 (15%)	91114.6/6.17	4507529	(U88878) Toll-like receptor 2 (*Homo sapiens*)
4	1.63×10^3	6/38 (15%)	56224.9/6.04	2501333	(U67513) acetolactate synthase, large subunit (*Methanococcus jannaschii*)

In view of these results, we next tested the approach with proteins of the bacterium *M. tuberculosis*. Despite the importance of this pathogen, relatively little is known about the bacterial proteins that are differentially expressed when it enters the host environment. This information is of some importance as it would provide valuable clues about its survival strategies in vivo as well as its mechanisms of pathogenesis.

Using a recently developed procedure (J. Dhillon *et al*., manuscript in preparation), *M. tuberculosis* cells were recovered from the lungs of infected mice and whole cell extracts prepared. Their patterns of protein expression were then compared with those of bacteria cultivated in vitro to different stages of growth. As expected relatively few cells of *M. tuberculosis* were recovered from animals and correspondingly few proteins could be recovered in amounts that allowed standard peptide mass fingerprinting (Figure 1). Using the orthogonal mass dataset approach, however, several polypeptides that were differentially up-regulated in the host were identified, e.g. succinyl-CoA synthetase (α subunit) and a hitherto hypothetical protein with similarity to enoyl-CoA hydratase. These results suggest the approach may be of some utility where limited peptides are available for analysis.

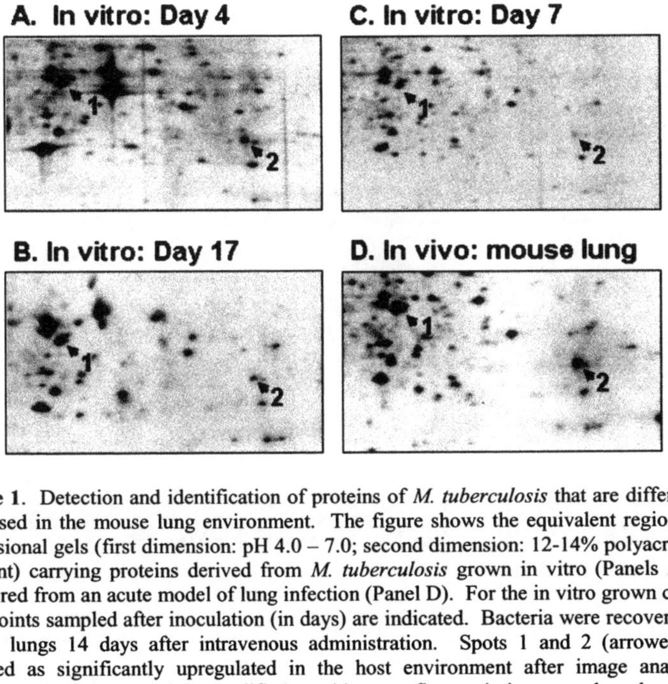

Figure 1. Detection and identification of proteins of *M. tuberculosis* that are differentially expressed in the mouse lung environment. The figure shows the equivalent regions of 2-dimensional gels (first dimension: pH 4.0 – 7.0; second dimension: 12-14% polyacrylamide gradient) carrying proteins derived from *M. tuberculosis* grown in vitro (Panels A-C) or recovered from an acute model of lung infection (Panel D). For the in vitro grown cells, the time points sampled after inoculation (in days) are indicated. Bacteria were recovered from mouse lungs 14 days after intravenous administration. Spots 1 and 2 (arrowed) were detected as significantly upregulated in the host environment after image analysis of replicate gels. The use of the modified peptide mass fingerprinting procedure described in the text identified these spots as succinyl-CoA synthetase (α subunit) and a putative enoyl-CoA hydratase. Further details will be published elsewhere.

Development of an enhanced DIOS chip

MALDI-MS is one of the most robust technologies for high throughput proteomics but is limited to the analysis of peptides of *m/z* >700. This is due to the need to add an ultraviolet light-absorbing chemical matrix that fragments during desorption, thereby creating molecular noise in the low end of the mass spectrum. Recent studies suggest that this problem can be avoided by direct desorption/ionisation of peptides into the gas phase from a porous silicon surface. This technology – termed DIOS – is a potentially important advance as it allows laser desorption/ionisation mass spectrometry to detect small peptides (and other molecules in this size range), thereby increasing the data available for protein identification searches using peptide mass fingerprinting [16]. However, factors such as silicon crystal orientation and surface area, etching conditions and the thermal conductivity of the surface all affect the efficiency of ion generation [17, 18]. There is therefore considerable variability in the performance characteristics of spongy silicon prepared by different protocols and significant scope for optimisation.

As an alternative, we have developed a novel type of DIOS chip with a radically different structure in the form of silicon columns having diameters in the nanometre range (Figures 2 and 3). These exhibit a high degree of uniformity in their heights, which can vary from sub-micron to several tens of microns by altering the fabrication parameters (M. Farooqui, unpublished results). The brush-like surface can be fabricated with high reproducibility and is confined to the target sample areas by use of a masking procedure. This allows the investigator to control the extent of sample penetration into the columnar structures and hence its degree of trapping of the substrate. This in turn affects the degree of accessibility to laser light during the desorption/ionisation process.

Although not yet tested, it is likely that optimal desorption and ionisation into the gas phase will depend on the shape and size of the molecule to be ionised as well as the dimensions of the columnar structure (column length, inter-column spacing etc.)

Figure 2. Photograph of an enhanced DIOS chip mounted on a standard MALDI sample plate. A one euro coin is also shown to give an indication of the scale. Due to the conductivity of the silicon substrate, no additional electrical connection is required between the chip and the metal carrier plate. The sample areas consist of columnar structures. This can be seen in greater detail in Figure 3.

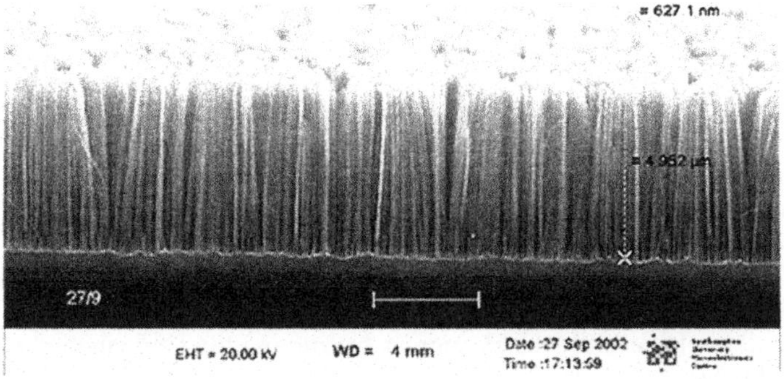

Figure 3. Scanning electron micrograph of a cross section through an enhanced DIOS chip, showing the regular brush-like columnar structures involved in sample trapping and desorption. In this case the columns are ~5 µm in length. However, the height of the silicon pillars is controllable and can be varied over a wide range to optimise sample trapping and ionisation. The scale bar is equivalent to 3 µm.

Since depth variation in the columnar structures affects the optical properties of the silicon surface, it can be readily monitored by reflectance spectroscopy, ellipsometry and imaging the visible photoluminescence under UV irradiation (Figure 4). To illustrate the ability of the enhanced DIOS chips in protein and peptide studies, Figure 5 shows a mass spectrum obtained for angiotensin I (molecular mass: 1296 Da). The results to date suggest that such peptides are desorbed into the gas phase with efficiencies that are at least as sensitive as conventional MALDI-MS. Moreover, negligible fragmentation is observed, which optimises the signal:noise ratio. It is concluded that the enhanced DIOS chips show considerable potential and merit further study to optimise sensitivity in proteomic studies.

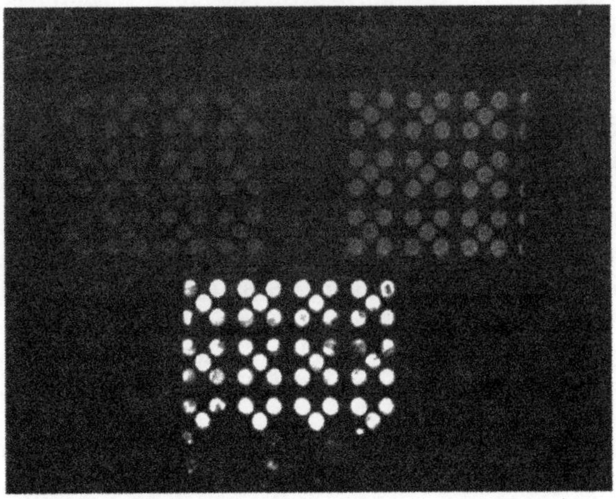

Figure 4. UV excited visible photoluminescence of the enhanced DIOS chips with different pillar heights. Since the optical properties of the sample target areas in each chip are dependent on the column length, UV light imaging can be used to determine the optimum length for ionisation of particular molecules. The dark spots in the active area are due to the analyte that has been dispensed.

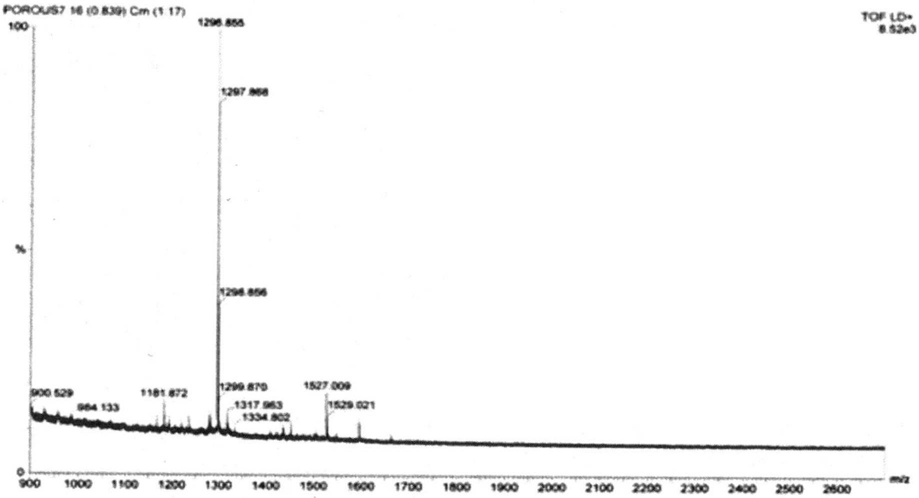

Figure 5. Example of a mass spectrum obtained using an enhanced DIOS chip. In this experiment, 1 μl of angiotensin I (1 nmol/ml in 0.1 % trifluoroacetic acid) was applied to the chip and 17 scans were acquired (*m/z* range 900-2,700) using a M@LDI mass spectrometer (Micromass). Data was processed using MassLynx 3.5 without background subtraction or calibration. The mass resolution was approximately 13000.

4. Discussion

Procedures such as MALDI-MS have revolutionised the throughput achievable in proteomics. However, protein identification still fails in a significant number of cases even when the protein in question is from an organism with a fully sequenced genome and theoretically contains sufficient information to assign it with high confidence. In this paper, we have described two approaches to deal with this problem and hence avoid the need to resort to slower, more labour-intensive procedures. The ability to successfully sample more of the expressed proteome in a high throughput fashion should significantly accelerate the discovery of novel biomarkers. It may also be of value in studies where only limited amounts of material are available for analysis.

Several other procedures have been reported to increase the information content of peptide mass spectra. One common approach is in vivo incorporation of specifically deuterated amino acid residues into proteins during cell growth in culture (see e.g. [19]). By comparing the abundance of the labelled peptides with their non-labelled counterparts derived from control cells, it is not only possible to obtain quantitative data on differentially expressed proteins but it should also be possible in principle to increase the accuracy of peptide mass fingerprinting. Clearly, this is a powerful approach but its use is restricted to the study of cultured cells.

In other cases, peptide fragments are derivatised after proteolysis of the protein of interest. For example, Brancia and co-workers recently reported the conversion of Lys residues to homoarginine moieties by guanidation [20]. While this has the advantage of identifying the C-terminal residue of a tryptic peptide and also improves the signal response, such peptides tend to fragment more readily, which disperses the signal. Additionally, the information content imparted by the derivatisation is not as extensive as that for methylation as only a single type of residue is affected, rather than the two (Asp and Glu) that are modified by methylation. It is also likely - but not proved - that methylation is less susceptible to steric hindrance, e.g. due to structural variations in a peptide, and hence may be a more efficient derivatisation than guanidation.

The other major drawback of using MALDI-MS for proteomics is the masking of signals from small peptides by the chemical noise generated by inclusion of a matrix molecule. The recent development of DIOS is a major step forward [21, 22]. However, the porous silicon surfaces that have been used to date display a significant degree of variability and are difficult to fabricate in a highly reproducible manner. The enhanced DIOS chips reported here should circumvent many of these problems, thereby increasing the amount of data that can be collected by a peptide mass fingerprinting. Among the other advantages of DIOS are accelerated sample preparation with consequent ease of automation, its greater tolerance to salt and its compatibility with existing MALDI platforms as well as with microfluidics and microchip technology. All of these advantages also apply to the enhanced DIOS chips reported here.

In many cases, the inclusion of data from just a single additional tryptic peptide can make the difference between a high confidence protein identification and a failed one. In more difficult cases, it should be possible to unite enhanced DIOS chips with the orthogonal mass dataset approach that is also described in this paper. The combination of the two promises to increase proteomic selectivity and sensitivity still further.

Acknowledgements

We thank Philip Butcher, Jas Dhillon and Denis Mitchison, St. George's Hospital Medical School, London for supplying the *M. tuberculosis* cells used in this study and for many useful discussions. These studies were supported by a pump-priming grant from the Faculty of Engineering & Applied Science, University of Southampton, by project grants from the BBSRC and MRC, and by an infrastructure award from the Science Research Investment Fund.

References

[1] L.A. Liotta, E.C. Kohn and E.F. Petricoin, Clinical proteomics: personalized molecular medicine. *JAMA* 286 (2001) 2211-2214.

[2] E.F. Petricoin, A.M. Ardekani, B.A. Hitt, P.J. Levine, V.A. Fusaro, S.M. Steinberg, G.B. Mills, C. Simone, D.A. Fishman, E.C. Kohn and L.A. Liotta, Use of proteomic patterns in serum to identify ovarian cancer. *Lancet* 359 (2002) 572-577.

[3] S. Kennedy, Proteomic profiling from human samples: the body fluid alternative. *Toxicol Lett* 120 (2001) 379-384.

[4] D.J.C. Pappin, P. Hojrup and A.J. Bleasby., Rapid identification of proteins by peptide mass fingerprinting. *Current Biology* 3 (1993) 327-332.

[5] W.J. Henzel, T.M. Billeci, J.T. Stults, S.C. Wong, C. Grimley and C. Watanabe, Identifying proteins from two-dimensional gels by molecular mass searching of peptide fragments in protein sequence databases. *Proc Natl Acad Sci U S A* 90 (1993) 5011-5015.

[6] J. Rosenfeld, J. Capdevielle, J.C. Guillemot and P. Ferrara, In-gel digestion of proteins for internal sequence analysis after one- or two-dimensional gel electrophoresis. *Anal Biochem* 203 (1992) 173-179.

[7] A. Shevchenko, I. Chernushevic, M. Wilm and M. Mann, "De novo" sequencing of peptides recovered from in-gel digested proteins by nanoelectrospray tandem mass spectrometry. *Mol Biotechnol* 20 (2002) 107-118.

[8] M. Mann and M. Wilm, Error-tolerant identification of peptides in sequence databases by peptide sequence tags. *Anal Chem* 66 (1994) 4390-4399.

[9] J. Wei, J.M. Buriak and G. Siuzdak, Desorption-ionization mass spectrometry on porous silicon. *Nature* 399 (1999) 243-246.

[10] A. Görg, C. Obermaier, G. Boguth, A. Harder, B. Scheibe, R. Wildgruber and W. Weiss, The current state of two-dimensional electrophoresis with immobilized pH gradients. *Electrophoresis* 21 (2000) 1037-1053.

[11] P. Adams, R. Fowler, G. Howell, N. Kinsella, P. Skipp, P. Coote and C.D. O'Connor, Defining protease specificity with proteomics: a protease with a dibasic amino acid recognition motif is regulated by a two-component signal transduction system in *Salmonella*. *Electrophoresis* 20 (1999) 2241-2247.

[12] S.-Y. Qi, A. Moir and C.D. O'Connor, Proteome of *Salmonella typhimurium* SL1344: identification of novel abundant cell envelope proteins and assignment to a two-dimensional reference map. *J Bacteriol* 178 (1996) 5032-5038.

[13] C.D. O'Connor, M. Farris, Hunt, L.G. and J.N. Wright, The proteome approach. In: P. Williams, J. Ketley, G.P.C. Salmond (eds.), Methods in Microbiology: Bacterial Pathogenesis. Academic Press, London, 2002, pp. 191-204

[14] A. Shevchenko, M. Wilm, O. Vorm and M. Mann, Mass spectrometric sequencing of proteins silver-stained polyacrylamide gels. *Anal Chem* 68 (1996) 850-858.

[15] M. Wilm, A. Shevchenko, T. Houthaeve, S. Breit, L. Schweigerer, T. Fotsis and M. Mann, Femtomole sequencing of proteins from polyacrylamide gels by nano-electrospray mass spectrometry. *Nature* 379 (1996) 466-469.

[16] J.J. Thomas, Z. Shen, J.E. Crowell, M.G. Finn and G. Siuzdak, Desorption/ionization on silicon (DIOS): a diverse mass spectrometry platform for protein characterization. *Proc Natl Acad Sci U S A* 98 (2001) 4932-4937.

[17] R.A. Kruse, X. Li, P.W. Bohn and J.V. Sweedler, Experimental factors controlling analyte ion generation in laser desorption/ionization mass spectrometry on porous silicon. *Anal Chem* 73 (2001) 3639-3645.

[18] Z. Shen, J.J. Thomas, C. Averbuj, K.M. Broo, M. Engelhard, J.E. Crowell, M.G. Finn and G. Siuzdak, Porous silicon as a versatile platform for laser desorption/ionization mass spectrometry. *Anal Chem* 73 (2001) 612-619.

[19] S.E. Ong, B. Blagoev, I. Kratchmarova, D.B. Kristensen, H. Steen, A. Pandey and M. Mann, Stable Isotope Labeling by Amino Acids in Cell Culture, SILAC, as a Simple and Accurate Approach to Expression Proteomics. *Mol Cell Proteomics* 1 (2002) 376-386.

[20] F.L. Brancia, A. Butt, R.J. Beynon, S.J. Hubbard, S.J. Gaskell and S.G. Oliver, A combination of chemical derivatisation and improved bioinformatic tools optimises protein identification for proteomics. *Electrophoresis* 22 (2001) 552-559.

[21] H. Föll, M. Christopherson, J. Constansen and G. Hasse, Formation and application of porous silicon. Mat. Sci and Engr. R 39 (2002) 93-141.

[22] W. Lewis, Z. Shen, M.G. Finn, and G. Siuzdak, Desorption-Ionization On Spectrometry (DIOS-MS): Background and Applications. *Int J Mass Spec* (2002) in press.

Understanding the Significance of Genetic Variability in the Human PON1 Gene

Clement E. FURLONG[1], Wan-Fen LI[2], Toby B. COLE[1,2], Rachel JAMPSA[1]
Rebecca J. RICHTER[1], Gail P. JARVIK[1,3], Diana M. SHIH[4], Aaron TWARD[4], Aldon J. LUSIS[4,5] and Lucio G. COSTA[2]
Departments of [1]Genome Sciences, Medicine (Division of Medical Genetics), [2]Environmental Health and [3]Epidemiology, University of Washington, Seattle, WA 98195, USA
Department of [4]Medicine, [5]Department of Microbiology, Immunology, and Molecular Genetics, UCLA, Los Angeles, CA 90095, USA

Abstract. A major goal of the environmental genome project has been to understand human genetic variability as it affects sensitivity or resistance to different environmental exposures. The human *PON1* gene was one of the early-identified genes whose polymorphisms affected both the detoxication of xenobiotics and metabolism of important physiological substrates. *PON1* encodes an HDL-associated enzyme that hydrolyzes a number of aromatic esters, organophosphorus (OP) compounds, drugs and oxidized lipid molecules. Our early studies demonstrated that of the two common coding region polymorphisms L55M and Q192R, it was the latter that affected the catalytic efficiency of hydrolysis of toxic organophosphorus (OP) insecticide metabolites. $PON1_{R192}$ hydrolyzes paraoxon and chlorpyrifos oxon with better catalytic efficiency than $PON1_{Q192}$ but soman and sarin with lower efficiency. Both 192 isoforms hydrolyze diazoxon and phenylacetate with approximately the same catalytic efficiency. While large differences in catalytic efficiency are observed for some substrates, the catalytic efficiency must reach a certain value to provide *in vivo* protection against OP exposures. In addition to the coding region polymorphisms that affect catalytic efficiency of substrate hydrolysis, polymorphisms that affect PON1 expression have been characterized in the 5' regulatory region of PON1. Experiments with PON1 knockout mice, PON1 injected knockout mice and transgenic mice expressing only human $PON1_{R192}$ or $PON1_{Q192}$ have provided important insights into the effects of both coding region and regulatory region polymorphisms on the modulation of OP exposures by the human plasma PON1 isoforms.

1. Introduction

The first observations of enzymatic hydrolysis of organophosphorus compounds by animal tissues are generally credited to Mazur [1]. These observations were followed by a series of carefully done studies by Aldridge [2,3] in which he observed that "A-esterases" could catalytically hydrolyze organophosphorus compounds including paraoxon (PO), whereas esterases that he categorized as "B-esterases" (e.g., carboxylesterases) were inhibited by organophosphorus compounds [2]. The plasma enzyme that hydrolyzed PO and arylesters was designated paraoxonase (PON1) or arylesterase (A-esterase)[1] [3].

Population distributions of PON1 activity in humans indicated a genetic polymorphism, with some individuals having high PON1 activity and others having low or intermediate activity. As noted below (Fig. 1), the methods used by these investigators could not have resolved bi- vs. trimodal distributions [4].

Cloning and characterization of the human PON1 cDNAs [5] led to the identification of the molecular basis of the PON1 activity polymorphism. Two coding region polymorphisms, L55M and Q192R, were identified [6,7]. The Q192R polymorphism affected rates of PO hydrolysis and, as described below, also the catalytic efficiency of hydrolysis of other substrates. Thus, it was thought that high PON1 activity would protect against parathion/paraoxon toxicity. Supporting this idea, species with high PON1 activity were more resistant to OP toxicity than were species with low activity [8-11].

2. Animal Models of Paraoxonase Variability

The first direct evidence that high PON1 levels were protective against OP exposure came from an experiment carried out by Main [12] in which he injected partially purified rabbit PON1 into rats and demonstrated increased resistance to PO exposure. Increasing the rats' plasma PON1 by 3- to 5-fold resulted in reduced mortality and prolonged time to death.

Main's experiment was followed up many years later by a series of experiments in our laboratory that were aimed at developing an animal model for examining the role of PON1 in OP detoxication. Our initial experiments involved the examination of the differential sensitivity of rabbits and rats to PO exposure using cholinesterase inhibition and mild symptoms as end points [10]. Rabbits, which have 7-fold higher plasma levels of PON1, were 4-fold less sensitive than rats to PO exposure via i.p. injection. Our next experiments, like Main's, involved injection of purified rabbit PON1 into rats and challenges with PO and chlorpyrifos oxon (CPO) via i.v., i.p., dermal or oral routes [13]. Injection of partially purified rabbit PON1 into the tail veins of rats resulted in a 9-fold increase in serum paraoxonase activity and a 50-fold increase in CPO activity. Protection against i.v. PO challenge was noted, but protection against PO via other routes of exposure was minimal. In contrast, injection of rabbit PON1 provided excellent protection against brain and diaphragm cholinesterase inhibition by CPO exposure via all routes of challenge. Our later results provided an explanation for these observations, as described below.

For the next series of experiments, we switched to mice as a model system [14]. Mice were chosen for two main reasons. Many more enzyme injection experiments could be carried out with a given amount of purified PON1, and genetics were much further advanced in mice, particularly the generation of knockout and transgenic animals.

The first series of experiments addressed the questions of the half-life of purified rabbit PON1 injected via different routes including i.v., i.p., i.v. plus i.p., and i.v. plus i.m., and the protection thus afforded by injected PON1 against exposure to CPO and its parent compound chlorpyrifos (CPS) [14]. As expected, injection of purified PON1 directly into the circulation produced an instant rise of plasma PON1 (Fig 1). The injected PON1 provided protection against both CPS and CPO. Further experiments demonstrated that injection of purified rabbit PON1 was protective against CPS and CPO toxicity when given pre-exposure (30 min or 24 h) or post exposure (30 min or 3h) [15]. These experiments provided convincing evidence that high levels of plasma PON1 are protective against exposure to specific OP compounds.

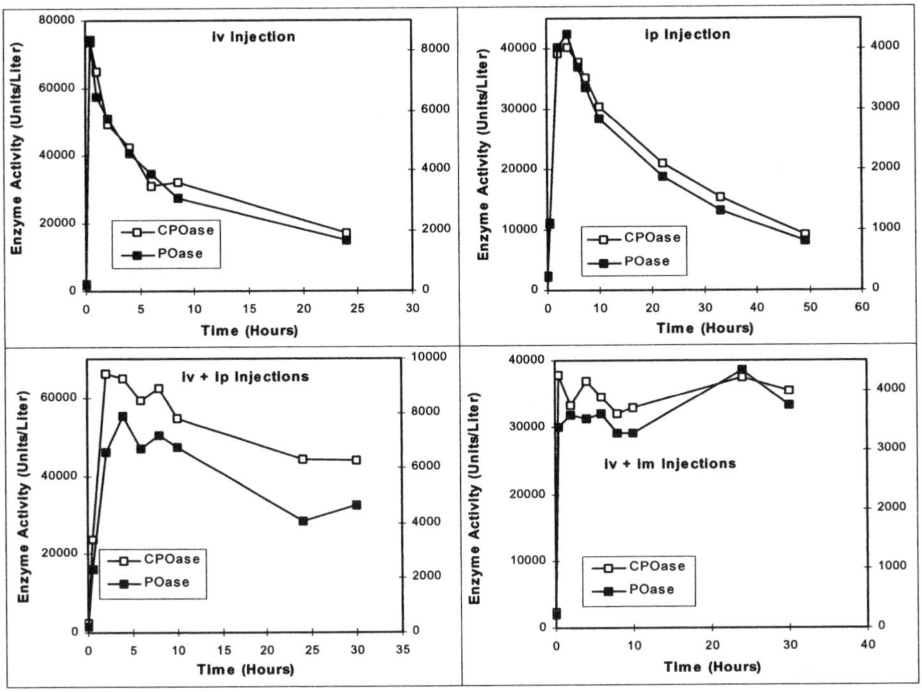

Figure 1. Time course of mouse serum paraoxonase activity following injection of rabbit paraoxonase into mice via the indicated routes. (From ref 14, with permission).

The generation of PON1 knockout mice by Shih et al. [16] added an additional important component to the mouse model. Disruption of the mouse *PON1* exon 1 resulted in the loss of both plasma [16] and liver PON1 (Fig. 2) [17]. As noted above, the PON1 injection experiments with wild type mice had provided convincing evidence that high levels of plasma PON1 are protective against CPO and CPS toxicity. The generation of the PON1 knockout mice allowed us to examine the effects of PON1 deficiency on sensitivity to specific OP compounds.

The PON1 knockout mice exhibited dramatically increased sensitivity to CPO [16] and DZO [18], but less to the respective parent organophosphorothioate compounds [16,18]. These mice, however, did not exhibit increased sensitivity to PO, the compound for which PON1 was named [18].

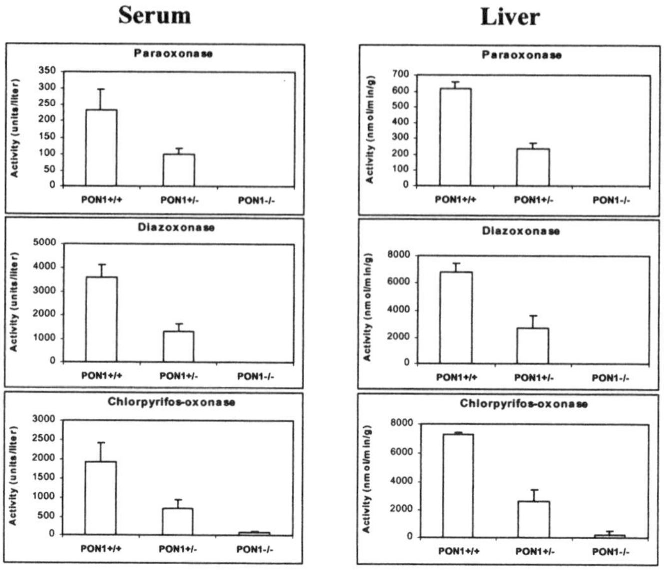

Figure 2. Activity levels of PON1 in wild-type mice (PON1$^{+/+}$), PON1 hemizygous mice (PON1$^{+/-}$) and PON1 knockout mice (PON1$^{-/-}$) for the hydrolysis of the indicated substrates.

The PON1 knockout mice provided another important addition to the mouse model system. Since they are devoid of both plasma and liver PON1, PON1 from any source can be injected (iv, ip or im) to restore plasma (and not liver) PON1 activity. The efficacy of the injected PON1 to provide protection against OP exposure could then be tested without any contribution from mouse PON1. Using this approach, Li et al. [18] found that either purified human PON1$_{192}$ isoform provided equal protection against DZO exposure, whereas PON1$_{R192}$ provided better protection against CPO exposure. Neither isoform protected against PO exposure. The explanation for these observations was provided by determining the catalytic efficiency of each purified human PON1$_{192}$ isoform [18]. The catalytic efficiencies of each PON$_{192}$ isoform for hydrolysis of DZO were identical, as was the ability of each injected isoform to protect against dermal exposure to DZO. The better protection of PON1$_{R192}$ against CPO exposure was matched with its better catalytic efficiency for hydrolyzing CPO. Even though PON1$_{R192}$ has a much higher efficiency for hydrolyzing PO than PON1$_{Q192}$, it is still approximately 11-fold lower than the efficiency of PON1 for hydrolyzing DZO. The lack of *in vivo* protection of PON1 against PO exposure had been predicted by Chambers and Pond [19]. Thus, it is the catalytic efficiency of a specific PON1 isoform that determines whether high levels of that PON1 can provide protection against a given OP exposure. The case of the human PON1$_{192}$ isoforms demonstrates that the change of a single amino acid can improve catalytic efficiency by as much as 9-fold. Based on the data available, it could be predicted that if the catalytic efficiency of PON1$_{R192}$ is increased by 10-fold, it would be a useful therapeutic for PO exposure. Interestingly, in our experiments where we injected purified rabbit PON1 (which hydrolyzes CPO very efficiently) into mice, we observed substantial protection against exposures to the parent compound CPS as well as CPO [14, 15]. These experiments provide important guidelines for generating recombinant forms of human PON1 that have catalytic efficiencies appropriate for specific therapeutic applications.

3. Genetic variability in the human *PON1* gene

The first DNA polymorphisms identified in the human PON1 gene were the two coding region polymorphisms L55M and Q192R [5] the latter of which affects the catalytic efficiency of hydrolysis of a number of substrates including the toxic metabolites of the insecticides [4,6,7], and some drugs [20]. $PON1_{Q192}$ has also been reported to protect against lipid oxidation better than $PON1_{R192}$ [21, 22]. The experiments described above have provided convincing evidence that both the PON1 Q192R polymorphism as well as PON1 levels contribute to resistance or sensitivity to specific OP exposures. The term PON1 status was introduced to include both position 192 genotype and PON1 levels, both of which are important in modulating OP exposures [14]. Other studies have shown that low PON1 status is a risk factor for cardiovascular disease [23, 24].

In attempts to understand the molecular basis of the large variability in PON1 levels among individuals (~13-fold), three research groups investigated polymorphisms in the PON1 5' regulatory region. Five polymorphisms were identified at positions -108, -126, -162, -832 and -909. Reporter gene assays provided information on the effects of each polymorphism on PON1 transcription and population studies complemented this information [25-29]. Some differences in position nomenclature related to base counts were noted among the three groups. The consensus of the three research groups was that the C-108 allele provided more efficient expression than the T-108 allele. Individuals homozygous for the C-108 allele have on average twice the level of plasma PON1 than individuals homozygous for the T-108 allele. Four polymorphisms were described in the 3' UTR of PON1, but were not tested for effects on PON1 expression [30]. The human *PON1* gene has 9 exons [31] and occupies approximately 26 kb on the long arm of chromosome 7. The *PON3* and *PON2* genes are immediately distal to *PON1*. Neither PON2 nor PON3 appear to hydrolyze OP compounds [32].

Recent determination of the sequence of the PON1 genes of 47 individuals revealed many previously undescribed polymorphisms [33]. Eight additional polymorphisms were found in the 5' regulatory region, 141 in the PON1 intronic sequences, a stop codon polymorphism in the coding region and 13 additional 3' UTR polymorphisms [33]. The termination codon polymorphism at position 194 does indeed result in inactivation of the $PON1_{R192}$ allele bearing this polymorphism [Jarvik et al., unpublished data]. The other recently discovered polymorphisms have yet to be characterized with respect to effects on PON1 levels. Analysis of the haplotypes inferred from this study [Jarvik et al. submitted] indicates that linkage disequilibrium within the PON1 gene extends over shorter blocks of sequence than observed in other regions of the genome [34-38]. However, significant linkage disequilibrium has been observed between $PON1_{R192}$ and $PON1_{L55}$ polymorphisms [and also between the $PON1_{T\text{-}108}$ and $PON1_{M55}$ polymorphisms [26].

4. Analysis of PON1 status

The data described above provide clear evidence that both PON1 levels and the Q192R polymorphism are important to consider in determining resistance to OP exposures. Eckerson et al. [39] provided the first approach for a two-substrate analysis of an individual's PON1 status. They plotted the rates at which plasma from individuals hydrolyzed PO vs. the rate at which they hydrolyzed phenylacetate. This approach provided a two-dimensional plot where the Q192R polymorphism as well as PON1 levels could be inferred. In a subsequent study, we plotted rates of hydrolysis of a number of different substrates against rates of PO hydrolysis [4]. Of all of these plots, only the plot of rates of diazoxon vs. PO hydrolysis provided a clear separation of all PON1 position 192 phenotpyes (QQ/QR/RR) [4, 23, 40]. This assay was converted to a high-throughput microtiter plate assay for large-scale studies [40]. PCR analysis verified the validity of this functional genomic analysis of PON1 status [23]. The enzyme analysis correctly phenotyped an individual whose *PON1*$_{R192}$ allele contained a termination codon at position 194 as noted above [Jarvik et al. unpublished data]. Figure 3 shows an example of this analysis.

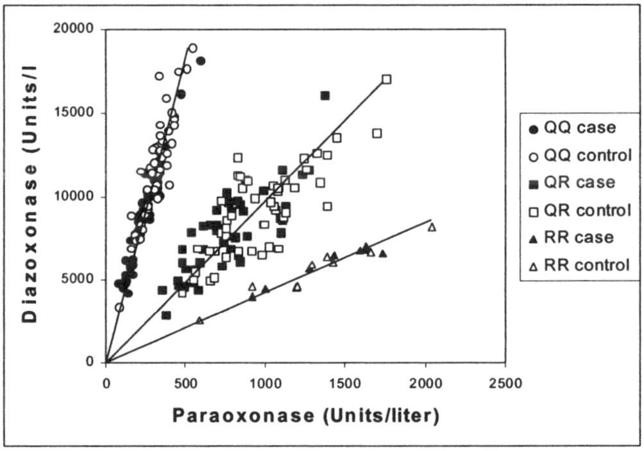

Figure 3. Two-dimensional analysis of PON1 status in 212 individuals (From ref. 23 with permission)

There are several points to note from Figure 3. First, the data points for the PON1$_{192}$ homozygotes (QQ and RR fall tightly along the trend lines, whereas some of the heterozytotes (QR vary considerably away from the trend line. The most likely explanation is that individuals to the left of the trend line are producing more PON1$_{Q192}$ while individuals to the right of the trend line are producing more PON1$_{R192}$. This seems to be a reasonable explanation since cis polymorphisms in the 5' regulatory region affect expression of PON1 as described above. Second, there is a large variability of PON1 levels among individuals of a given PON1 inferred genotype as a result of both regulatory region polymorphisms as well as probable contributions from differences in genetic background. Third, this plot accurately infers position 192 genotype. The genotypes in the figure were assigned by PCR analysis [23]. As noted above, this is a functional genomic analysis, since this procedure is capable of detecting a non-functional allele in heterozygotes.

Following our description of polymerase chain reaction (PCR) analytical procedures for characterizing the L55M and Q192R polymorphisms, more than 50 studies have been carried out examining possible relationships between PON1 genotype(s) and disease states [reviewed in 30]. These studies considered only the *PON1* (and in some cases *PON2* and *PON3*) genotype(s) despite the large, well-known variability in PON1 levels among individuals [4, 23, 40]. Some of the more recent studies have considered PON1 levels [23, 24, 29, 41]. It should be clear to anyone with a fundamental appreciation of biochemistry that an individual with plasma PON1 levels that detoxicate an OP or oxidized lipid at 10-times or greater the rate of another individual, should be significantly more resistant to the effects of the toxicant. All of the data described above support this basic concept. While PCR methodology is tempting to apply to epidemiological studies, the case of PON1 status where a high throughput functional assay is available that provides significantly more information, it doesn't make sense to carry out only genotyping assays. One case where an extensive knowledge of genetic haplotypes would be informative is where the actual disease linked gene is in linkage disequilibrium with the *PON1* gene. The haplotype data available to date [33] do not suggest that this will be the case, however, long-range haplotype data that extend into the *PON2* and *PON3* genes and upstream from *PON1* are not yet available. In many studies, plasma samples were collected in EDTA, which irreversibly inactivates PON1, preventing assessment of PON1 status. The importance of considering PON1 levels for epidemiological studies is such that it may well be worth making use of immunological methods [42] for estimating PON1 levels where frozen plasma samples are still available.

5. Possible therapeutic uses of recombinant PON1

The current protocol for treating cases of poisoning by OPs such as nerve agents and insecticides involves administration of atropine to block acetylcholine receptors and PAM to regenerate inhibited cholinesterase [43]. Ideally, treatment would also include an efficient enzyme that would rapidly and catalytically inactivate the OP compound. One of the best candidate enzymes for treating individuals exposed to OP compounds is recombinant human PON1. The toxicology studies described above indicate that it will be important to generate variants of PON1 with catalytic efficiencies tailored to specific OP compounds. Where the specific OP is unknown, it may be useful to have a cocktail of different variants of PON1 available. Our recent expression of active human PON1 in *E. coli* li makes it possible to engineer variants of PON1 with higher catalytic efficiencies for specific OP compounds (Jampsa et al., unpublished data). The data described above indicate that recombinant $PON1_{R192}$ or plasma from an individual with very high levels of $PON1_{R192}$ may be useful for treating CPO exposures. Either $PON1_{192}$ isoform or plasma from individuals expressing very high PON1 levels should be useful for treating diazoxon exposures. Treatment of nerve agent exposure would require PON1 variants with higher catalytic efficiencies [18]. Rabbit PON1, which is more stable than human PON1 [44], is an excellent candidate to engineer for decontamination applications.

6. Research needs

To date, there are few epidemiological studies examining the relationship between PON1 status and susceptibility to OP toxicity. The information necessary for an informative study would include the nature and level of exposure, the consequence(s) of the exposure (e.g., cholinesterase inhibition) and the subject's PON1 status.

PON1 status is important for modulating exposure to the oxon forms of diazinon and CPS. It is clear that many exposures include significant oxon residues [45]. Since the oxon forms can inhibit cholinesterase at a rate 1000 times faster than parent compounds [46], a small percentage of oxon is very significant as is an individual's PON1 status in modulating these exposures. PON1 appears to play less of a role in modulating exposure to the parent compounds [16, 18]. Cytochromes P450 appear to play the major role in detoxifying as well as activating the parent organophosphorothioate compounds [47]. Research is beginning to elucidate the specific cytochromes P450 that metabolize the organophosphorus insecticides [48]. However, animal models that will assist in examining human genetic variability of the cytochrome P450 systems in OP metabolism are not yet established.

Currently, cholinesterase inhibition is used as an endpoint to assess OP exposure [49]. Recent reports note that other important targets are modified by OP exposures [46, 50-52]. The availability of microarray analytical procedures will provide investigators with the tools necessary to identify cellular pathways that are disrupted by OP exposure. Evaluation of data from such experiments will provide a better basis for establishing exposure guidelines.

As noted above, PON1 and recombinant variant PON1s have an excellent potential for treating OP exposures. Research in this area should provide fruitful results.

7. Summary

The human PON1 gene exhibits a number of polymorphisms in the 5'regulatory region, the coding sequence, intervening sequences and in the 3' UTR. At least one of the 5' regulatory region polymorphisms (C-108T) has a significant influence on levels of plasma PON1. The Q192R coding region polymorphism influences the catalytic efficiency of hydrolysis of a number of PON1 substrates. The W194X polymorphism results in the truncation and inactivation of the allele bearing the termination codon [Jarvik et al. submitted]. It is the combination of the Q192R polymorphism plus the level of PON1 expression that determines an individual's capability for detoxifying compounds processed by PON1, including specific OP compounds and oxidized lipids as well as rates of drug metabolism. The recently revealed polymorphisms in the 5' regulatory region and PON1 introns [33] have not yet been characterized with respect to affects on PON1 expression. The "humanized" mice in which mouse PON1 has been replaced with either human $PON1_{Q192}$ or $PON1_{R192}$ have provided an excellent system for evaluating the function of these human PON1 isoforms under physiological conditions. The results of the toxicology experiments with these mice have pointed out the higher risk of CPO exposure associated with the $PON1_{Q192}$ allele. The knockout mice have illustrated the importance of PON1 levels in modulating exposures to diazoxon and chlorpyrifos oxon. Together, these experiments have pointed out the usefulness of knockout and "humanized" mice in understanding genetic variability and sensitivity to OP exposures.

The recent sequencing of the complete *PON1* gene from 47 individuals illustrates the capabilities of high-throughput genomic analysis for identifying genetic variability among individuals. The development of the mouse model system and approaches for studying the functionality of the regulatory region polymorphisms illustrate approaches that are useful in understanding the functional consequences of the observed genetic variation. The lesson learned from the study of the genetic variability of PON1 is that both the quality and quantity of PON1 are important in modulating exposure to environmental toxicants and drugs, as well as endogenously generated toxic lipid molecules.

Notes

1. While earlier publications used the terms paraoxonase, arylesterase or A-esterase to describe paraoxonase activity, we will use PON1 throughout this text.

References

[1] A. Mazur, An Enzyme in Animal Tissue Capable Of Hydrolyzing the Phosphorus-Fluorine Bond of Alkyl Fluorophosphates, *J. Biol Chem.* **146** (1946) 271-289.

[2] W.N. Aldridge, Serum Esterases 1. Two Types Of Esterase (A and B) Hydrolyzing P-Nitrophenyl Acetate, Propionate and Butyrate and a Method for Their Determination, *Biochem. J.* **53** (1953) 110-117.

[3] W.N. Aldridge, An Enzyme Hydrolysing Diethyl-P-Nitrophenyl Phosphate (E-600) and Its Identity with the A-Esterase Of Mammalian Sera, *Biochem. J.* **53** (1953) 117-124.

[4] H. Davies, R.J. Richter, M. Keifer, C. Broomfield, J. Sowalla and C.E. Furlong, The Effect of the Human Serum Paraoxonase Polymorphism is Reversed with Diazoxon, Soman and Sarin, *Nat. Genet.* **14** (1996) 334-336.

[5] C. Hassett, R.J. Richter, R. Humbert, C. Chapline, J.W. Crabb, C.J. Omiecinski and C.E. Furlong, Characterization Of cDNA Clones Encoding Rabbit and Human Serum Paraoxonase: The Mature Protein Retains Its Signal Sequence, *Biochemistry* **30** (1991) 10141-10149.

[6] R. Humbert, D.A. Adler, C.M. Disteche, C. Hassett, C.J. Omiecinski and C.E. Furlong, The Molecular Basis of the Human Serum Paraoxonase Activity Polymorphism, *Nat. Genet.* **3** (1993) 73-76.

[7] S. Adkins, K. N. Gan, M. Mody, B. N. La Du, Molecular Basis for the Polymorphic Forms of Human Serum Paraoxonase/Arylesterase: Glutamine or Arginine At Position 191, for the Respective A or B Allozymes, *Am. J. Hum. Genet.* **52** (1993) 598-608.

[8] S. B. McCollister, R. J. Kociba, C. G. Humiston and D. D. McCollister, Studies on the Acute and Long-term Oral Toxicity of Chlorpyrifos (0,0- diethyl-0(3,5,6-trichloro-2-pyridyl) phosphorothioate), *Food Cosmet. Toxicol.* **12** (1974) 45-61.

[9] C.J. Brealey, C.H. Walker and B.C. Baldwin, A-Esterase Activities in Relation to the Differential Toxicity of Pirimiphos-Methyl to Birds and Mammals, *Pest. Sci.* **11** (1980) 546-554.

[10] L.G. Costa, R.J. Richter, S.D. Murphy, G.S. Omenn, A.G. Motulsky and C.E. Furlong, Species Differences in Serum Paraoxonase Activity Correlate with Sensitivity to Paraoxon Toxicity. In: Nato ASI Series, Vol. H13. Toxicology of Pesticides: Experimental, Clinical and Regulatory Aspects. L.G. Costa, et al. (eds), Springer-Verlag, Berlin, Heidelberg 1987, pp. 263-266.

[11] C.E. Furlong, R.J. Richter, S.L. Seidel, L.G. Costa and A.G. Motulsky, Spectrophotometric Assays For The Enzymatic Hydrolysis of The Active Metabolites Of Chlorpyrifos and Parathion by Plasma Paraoxonase/ Arylesterase, *Anal. Biochem.* **180** (1989) 242-247.

[12] A.R. Main, The Role of A-Esterase in The Acute Toxicity of Paraoxon, TEPP and Parathion, *Can. J. Biochem. Physiol.* 34 (1956) 197-216.

[13] L.G. Costa, B.E. McDonald, S.D. Murphy, G.S. Omenn, R.J. Richter, A.G. Motulsky and C.E. Furlong, Serum Paraoxonase and Its Influence on Paraoxon and Chlorpyrifos-Oxon Toxicity in Rats, *Toxicol. Appl. Pharmacol.* **103** (1990) 66-76.

[14] W.-F. Li, L.G. Costa and C.E. Furlong, Serum Paraoxonase Status: A Major Factor In Determining Resistance to Organophosphates, *J. Toxicol. Environ. Health* **40** (1993) 337-346.

[15] W.-F. Li, C.E. Furlong and L.G. Costa, Paraoxonase Protects Against Chlorpyrifos Toxicity in Mice, *Toxicol. Lett.* **76** (1995) 219-226.

[16] D.M. Shih, L. Gu, Y.-R. Xia, M. Navab, W.-F. Li, S. Hama, L.W. Castellani, C.E. Furlong, L.G. Costa, A.M. Fogelman and A.J. Lusis, Mice Lacking Serum Paraoxonase Are Susceptible to Organophosphate Toxicity and Athersclerosis, *Nature* **394** (1998) 284-287.

[17] C.E. Furlong, W-F Li, V.H. Brophy, G.P. Jarvik, R.J. Richter, D.M. Shih, A.J. Lusis and L.G. Costa, The *PON1* Gene and Detoxication, *NeuroToxicol.* **21** (2000) 581-588.

[18] W.-F. Li, L.G. Costa, R.J. Richter, T. Hagen, D.M. Shih, A. Tward, A.J. Lusis and C.E. Furlong, Catalytic efficiency determines the *in vivo* efficacy of PON1 for detoxifying organophosphates, *Pharmacogenetics* **10** (2000) 767-780.

[19] A.L. Pond, H.W. Chambers, J.E. Chambers, Organophosphate Detoxication Potential of Various Rat Tissues via A-Esterase and Aliesterase Activities, *Toxicol. Lett.* 78 (1995) 245-252.

[20] S. Billecke, D. Dragonov, R. Counsell, P. Stetson, C. Watson, C. Hsu, B.N. La Du, Human serum paraoxonase (PON1) isozymes Q and R hydrolyze lactones and cyclic carbonate esters, Drug Metab Dispos **28** (2000) 1335-1342.

[21] M. Aviram, E. Hardak, J. Vaya, S. Mahmood, S. Milo, A. Hoffman, S. Billicke, D. Draganov and M.Rosenblat, Human Serum Paraoxonases (PON1) Q And R Selectively Decrease Lipid Peroxides in Human Coronary and Carotid Atherosclerotic Lesions: PON1 Esterase and Peroxidase-like Activities, *Circulation* **101** (2000) 2510-2517.

[22] B. Mackness, M.I. Mackness, S. Arrol, W. Turkie, P.N. Durrington, Effect of the Human Serum Paraoxonase 55 and 192 Genetic Polymorphisms on the Protection by High Density Lipoprotein Against Low Density Lipoprotein Oxidative Modification, *FEBS Lett.* **423** (1998) 57-60.

[23] G.P. Jarvik, L.S. Rozek, V.H. Brophy, T.S. Hatsukami, R.J. Richter, G.D. Schellenberg, and C.E. Furlong, Paraoxonase Phenotype Is a Better Predictor of Vascular Disease than $PON1_{192}$ or $PON1_{55}$ Genotype, *Atheroscler. Thromb. Vasc. Biol.* **20** (2000) 2442-2447.
[24] B. Mackness, K. D. Gershan, et al. Paraoxonase Status in Coronary Heart Disease: Are Activity and Concentration More Important Than Genotype?" *Arterioscler. Thromb. Vasc. Biol.* **21** (2001) 1451-1457.
[25] V.H. Brophy, M.D. Hastings, J.B. Clendennning, R.J. Richter, G.P. Jarvik and C.E. Furlong, Polymorphisms in the Human Paraoxonase (*PON1*) Promoter, *Pharmacogenetics* **11** (2001) 77-84.
[26] V.H. Brophy, R.L. Jampsa, J.B. Clendenning, L.A. McKinstry, G.P. Jarvik and C.E. Furlong, Effects of 5' regulatory region polymorphisms on paraoxonase (PON1) expression, *Am. J. Hum. Genet.* **68** (2001) 1428-1436.
[27] I. Leviev, RW James, Promoter Polymorphisms of Human Paraoxonase PON1 Gene and Serum Paraoxonase Activities and Concentrations, *Arterioscler. Thromb. Vasc. Biol.* **20** (2000) 516-521.
[28] T. Suehiro, T. Nakamura, M. Inoue, T. Shiinoki, Y. Ikeda, Y. Kumon, M. Shindo, H Tanaka and K. Hashimoto A polymorphism upstream from the human paraoxonase (PON1) gene and its association with PON1 expression, *Atherosclerosis* **150** (2000) 295-298.
[29] R.W. James, I. Leviev, J. Ruiz, P. Passa, P. Froguel and M.-C. Blatter Garin, Promoter Polymorphism T(-107)C of the Paraoxonase (PON1) Gene Is a Risk Factor for Coronary Heart Disease in Type 2 Diabetic Patients, *Diabetes* 49 (2000) 1390-1393.
[30] V.H. Brophy, G.P. Jarvik and C.E. Furlong, PON1 Polymorphisms. In: Paraoxonase (PON1) in Health and Disease: Basic and Clinical Aspects. L.G. Costa and C.E. Furlong (eds.), Kluwer Academic Press, Boston, 2002, pp. 53-77.
[31] J.B Clendenning, R. Humbert, E.D. Green, C. Wood, D. Traver and C.E. Furlong, Structural Organization of the Human PON1 Gene, *Genomics* **35** (1996) 586-589.
[32] S.L. Primo-Parmo, R.C. Sorenson, J. Teiber, B.N. La Du, The Human Serum Paraoxonase/Arylesterase Gene (PON1) Is One Member Of a Multigene Family *Genomics* **33** (1996) 498-507.
[33] *SeattleSNPs. NHLBI Program for Genomic Applications, UW-FHCRC, Seattle, WA (URL: http://pga.mbt.washington.edu) [November, 2002].*
[34] M.J. Daly, J.D. Rioux, S.F. Schaffner, T.J. Hudson, E.S. Lander, High-Resolution Haplotype Structure In The Human Genome, *Nat. Genet.* **29** (2001) 229-232.
[35] D.B. Goldstein: Islands of Linkage Disequilibrium, *Nat. Genet.* **29** (2001) 109-111.
[36] A.J. Jeffreys, L. Kauppi, R. Neumann, Intensely Punctate Meiotic Recombination in the Class II Region of the Major Histocompatibility Complex, *Nat. Genet.* **29** (2001) 217-222.
[37] G.C. Johnson, L. Esposito, B.J. Barratt, *et al.* Haplotype Tagging for the Identification of Common Disease Genes, *Nat. Genet.* **29** (2001) 233-237.
[38] N. Patil, A.J. Berno, D.A. Hinds, *et al.* Blocks Of Limited Haplotype Diversity Revealed by High-Resolution Scanning of Human Chromosome 21, *Science* **294** (2001) 1719-1723.
[39] H.W. Eckerson, C.M. Wyte and B.N. La Du, The Human Serum Paraoxonase/Arylesterase Polymorphism. Am. J. Hum. Genet. **35** (1983) 1126-1138.
[40] R.J. Richter, and C.E. Furlong, Determination Of Paraoxonase (PON1) Status Requires More Than Genotyping, *Pharmacogenetics* **9** (1999) 745-753.
[41] M.I. Mackness, P.N. Durrington, A. Ayub, B. Mackness. Low Serum Paraoxonase: A Risk Factor for Atherosclerotic Disease? *Chem. Biol. Interact.* 119-120 (1999) 389-397.
[42] M.C. Blatter Garin, C. Abbott, S. Messmer, M. Mackness, P. Durrington, D. Pometta, R.W. James, Quantification of Human Serum Paraoxonase by Enzyme-linked Immunoassay: Population Differences in Protein Concentrations, *Biochem.J.* **304** (1994) 549-554.
[43] National Research Council.. *Chemical and Biological Terrorism: Research and Development to Improve Civilian Medical Response*. National Academy Press. Washington, D.C. 1999.
[44] C.E. Furlong, R.J. Richter, C. Chapline and J.W. Crabb, Purification of Rabbit and Human Serum Paraoxonase, *Biochemistry* **30** (1991) 10133-10140.
[45] K.L. Yuknavage, R.A. Fenske, D.A. Kalman, M. C. Keifer and C.E. Furlong, Simulated Dermal Contamination with Capillary Samples and Field Cholinesterase Biomonitoring, *J. Toxicol. and Env. Health* **51** (1997) 35-55.
[46] R.A. Huff, J.J. Corcoran, J.K. Anderson, M.B. Abou-Donia, Chlorpyrifos Oxon Binds Directly to Muscarinic Receptors and Inhibits Camp Accumulation in Rat Striatum, *J. Pharmacol. Exp. Ther.* **269** (1994) 329-335.
[47] L.G. Sultatos, L.D. Minor, S.D. Murphy, Metabolic Activation of Phosphorothioate Pesticides: Role of the Liver, *J. Pharmacol. Exp. Ther.* **232** (1985) 624-628.
[48] D. Dai, J. Tang, R. Rose, E. Hodgson, R.J. Bienstock, H.W. Mohrenweiser and J.A. Goldstein, Identification of Variants of CYP3A4 and Characterization of Their Abilities to Metabolize Testosterone and Chlorpyrifos, *J. Pharmacol. Exp. Ther.* **299** (2001) 825-31.

[49] C. Timchalk, R.J. Nolan, A.L. Mendrala, D.A. Dittenber, K.A. Brzak and J.L. Mattsson, A Physiologically Based Pharmacokinetic and Pharmacodynamic (PBPK/PD) Model for the Organophosphate Insecticide Chlorpyrifos in Rats and Humans, *Toxicol. Sci.* **66** (2002) 34-53.

[50] S.J. Garcia, F.J. Seidler, D. Qiao and T.A. Slotkin. Chlorpyrifos Targets Developing Glia: Effects On Glial Fibrillary Acidic Protein, *Brain Res. Dev.* **133** (2002) 151-61.

[51] J.A. Bomser and J.E. Casida. Diethylphosphorylation of Rat Cardiac M2 Muscarinic Receptor by Chlorpyrifos Oxon in Vitro, *Toxicol. Lett.* **119** (2001) 21-26.

[52] P.G. Richards, M.K. Johnson and D.E. Ray, Identification of Acylpeptide Hydrolase as A Sensitive Site for Reaction with Organophosphorus Compounds and A Potential Target for Cognitive Enhancing Drugs, *Mol. Pharmacol.* **58** (2000) 577-583.

Functional Genomics Methods in Hepatotoxicity

Wilbert H.M. HEIJNE*, Rob H. STIERUM, Robert-Jan A.N. LAMERS and Ben van OMMEN
TNO Nutrition and Food Research, PO box 360, 3700 AJ Zeist, The Netherlands
** Corresponding author: Phone: +31 30 694 4137; Fax: + 31 30 696 02 64*
E-mail address: Heijne@voeding.tno.nl

Abstract. Functional genomics technologies including genomics, transcriptomics, proteomics and metabonomics prove to be of great value in life sciences. Application of these technologies in toxicology is discussed and illustrated with examples. Toxicogenomics involves the integration of conventional toxicological examinations with patterns of gene or protein expression or metabolite profiles. The expectations are earlier and more sensitive detection of toxicity as well as the possibility to elucidate mechanisms at the molecular level rather than making empirical observations. A combined transcriptomics and proteomics measurement of bromobenzene hepatotoxicity in rats is presented, where also the individual urine metabolite profiles are measured and compared using pattern recognition techniques. Both known and yet unknown effects in bromobenzene-induced liver toxicity were identified. Secondly, a study of the hepatotoxicity of food additives shows that after treatment with these compounds that show only moderate toxicity, transcriptomics measurements are able to point out relevant information on the mechanism of action.

1. Introduction

1.1 Developments in toxicology

Toxic effects of substances in the environment on an organism have been studied for many years, initially only by looking at the morphology and physiology of the organism, including lethal dose determination, body and organ weight, and gross pathology. The development of histopathological techniques enabled the microscopic examination of tissues and the determination of toxic effects at the cellular level. Gradually, toxicologists developed assays for sensitive and early identification of specific endpoints and for discrimination between different types of toxicity. This results in a thereby shortening of the time from administration to detection of effects and thus the animal exposure. Changes in levels of particular proteins or metabolites in tissue, blood or urine that correlate well with certain types of toxicity are now routinely assessed. Functional genomics technologies have emerged and provide toxicologists with new possibilities to study toxicity in an organism at the molecular level in a cell-wide approach.

1.2 Functional genomics

Novel functional genomics technologies measure cellular responses in an organism in a holistic way. The knowledge of the genome sequence and the protein and metabolite contents of a cell is used to deduce functional roles in mechanisms and cellular pathways of the different biomolecules. The area of research that studies the genome through analysis of the nucleotide sequence, the genome structure and its composition is called genomics. Determination of expression levels of gene transcripts, proteins or metabolites were named transcriptomics, proteomics and metabonomics, respectively.

Genomics and transcriptomics - The composition and organization of the genome determines many biological processes in an organisms thereby influencing the susceptibility towards genetic diseases or the response to xenobiotic compounds. Many cellular processes are controlled at level of the gene expression. Gene expression measurement by determination of the mRNA levels has shown to be valuable in the prediction of protein synthesis and protein activity (e.g., the total enzyme activity towards a substrate). Initially, mRNA levels were determined using Northern blots or RT-PCR (Reverse transcription polymerase chain reaction). cDNA microarrays were developed as a large-scale method for gene expression measurement which takes advantage of the availability of collections (libraries) of gene fragments with a known sequence and often with annotation on its (putative) function in the cell. As in Northern blots, the specific hybridization capacity of single stranded DNA and RNA is used to determine specific mRNA levels of the gene of interest. However, instead of one gene at the time, single-strand cDNA molecules for thousands of different genes are deposited each in a fixed spot on a surface (e.g., glass slide, plastic, nylon membrane), a cDNA microarray or DNA chip. By hybridization of the cDNA micorarray with a pool of isolated mRNAs, each specific mRNA will only hybridize with the cDNA in the spot containing the complementary cDNA for this specific gene. The amount of cDNA hybridized in each spot can be detected, by measurement of fluorescence that was incorporated in the sample. An overview of the process of cDNA microarray-based mRNA level measurements is shown in figure 1. In practice, always two samples are hybridized together on the cDNA microarray, where one test sample is labeled with one type of fluorophore (e.g., green fluorescent), and one reference or control sample is labeled with another type (e.g., red fluorescent). Quantification of the amount of both types of fluorescence enables the determination of a ratio of expression for each gene in the test sample with respect to the control sample. Whereas the large part of the thousands of genes measured will not show changes in expression level when controls are compared to diseased or treated samples, the genes that are found to be induced or repressed provide a wealth of information on cellular mechanisms affected at the gene expression level by the disease or treatment. For more in-depth reading on cDNA microarrays technology, we refer to references [1] and [2].

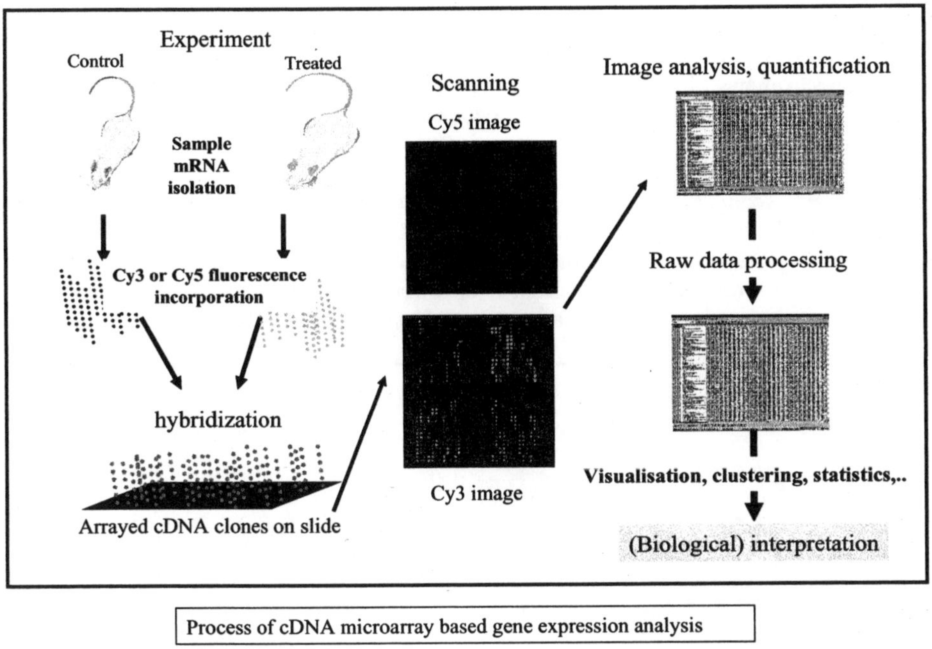

***Figure 1** : Process of transcriptomics measurement using cDNA microarrays*

Proteomics – It is highly likely that not all cellular mechanisms can be identified at the mRNA level. Especially cell-protective mechanisms of response might require fast modification or subcellular redistribution of proteins already present in the cell. Protein analysis technologies may have a better chance to visualize processes that do not involve active biosynthesis, or at least they will be complementary to gene expression analysis. The term "proteome" describes the complete population of expressed proteins in a cell. In analogy to transcriptomics, proteomics technologies are applied for the simultaneous measurement of the thousands of proteins in a cell. Proteomics will be of great benefit for toxicology, or even more important than measurement of the transcriptome, since proteins predominantly act in the cellular reactions rather than the gene transcripts. The measurement of the proteome is more complex than transcriptomics, where the total mRNA population can be isolated at once and, relatively easy, the thousands of transcripts that only differ in nucleotide sequence are sorted in the process of cDNA microarray hybridization. In contrast, to separate all different proteins from a cell is a very complicated task, as all proteins have different properties (mass, isoelectric point, solubility, stability, etc). Also, post-translational modifications and the capability to form protein complexes complicate the measurement of. Various methods have been developed in proteomics research. One method combines the relatively old technique of two-dimensional agarose gel electrophoresis with the developments of powerful automated image analysis software, advances in mass spectrometry and internet-based global information exchange. The total protein content of a sample is first separated on the basis of the isoelectric point (pI) of the proteins, using a strip with an immobilized pH gradient. Different pH gradients can be chosen in order to obtain maximal separation in the area of interest. After isoelectric focusing, proteins are transferred to a polyacrylamide gel and based on protein mass, separation is effected by standard gel electrophoresis. Separated proteins are then visualized using fluorescent or silver staining. Gels are scanned and images are analyzed with dedicated software and spot volumes are quantified. Characteristic gel images of protein patterns obtained from liver of control and bromobenzene-treated rats are shown in figure 2. Subsequently, spots of interest can be isolated from the gel, purified and analyzed by means of mass spectrometry. Proteins can be identified after a specific fragmentation (e.g., digestion with trypsin), which generates peptide fragments with a very specific mass determined by mass spectrometry. The fragmentation pattern is then used to identify what protein matches this pattern in a database search with predicted fragment patterns of all known protein sequences. Also, peptide sequencing can be conducted by MS/MS techniques to further confirm protein identities. Besides actual isolation, protein spots can be putatively identified based on a match with a previously identified protein in the exact same position on a reference gel. Though with lower accuracy, spots can be identified by prediction of their position on the gel after calculation of mass and isoelectric point. [3]

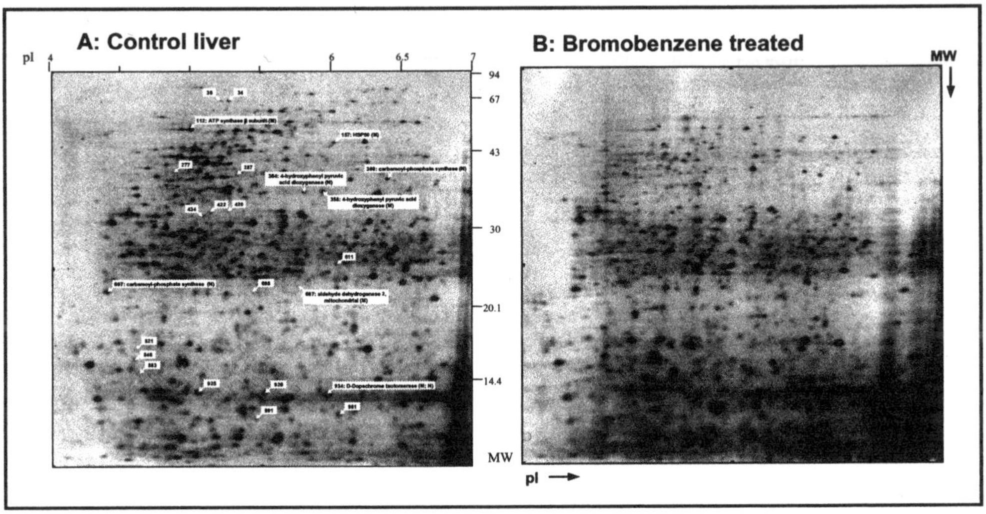

Figure 2.A. *The 2D-gel images with the protein pattern obtained from an untreated animal is shown. Spot numbers of proteins that were differentially expressed in bromobenzene or corn oil treated animals are indicated.* ***Figure 2. B*** *shows a typical gel image of rat liver proteins after bromobenzene treatment.*

Metabonomics - The effect of cellular processes is reflected in the metabolite levels, which can be measured in the cell and in the extracellular fluids. This provides researchers with a major advantage, as non-invasive methods can be used to collect samples such as blood (plasma) urine, or other body fluids, largely increasing the applicability in both human and animal experiments. Mass spectrometric techniques (GC-MS) allow low concentration components to be measured individually. For global screening, ^{1}H Nuclear Magnetic Resonance spectroscopy (NMR) is an attractive approach, as a wide range of metabolites can be quantified at the same time without extensive sample preparation. A spectrum is obtained with resonance peaks characteristic for all small biomolecules. In this way, a metabolic fingerprint is obtained characterizing the biological fluid under study. In the spectrum, individual signals can be quantified and identified based on available reference spectra. NMR spectra of biological fluids are very complex due to the mixture of numerous metabolites present in these fluids. Variations between samples are often too small to be recognized by eye. In order to increase the comparability of NMR spectra and thereby maximize the power of the subsequent data analysis, a partial linear fit algorithm was developed which adjusts minor shifts in the spectra while maintaining the resolution. To find significant differences, multivariate data analysis is needed to explore recurrent patterns in a number of NMR spectra. In figure 3, a factor spectrum is used to identify the metabolite NMR peaks that differ in rat urine upon bromobenzene treatment compared to controls. Correlation between variables in the complex and large data sets (thousands of signals in spectra) are related to a target variable such as toxicity status. The combination of mass spectrometric profiling with multivariate data analysis provides a powerful fingerprinting methodology. For an exhaustive analysis of a complex mixture of metabolites, a combination of analytical techniques is desirable. [4] [5;6;7]

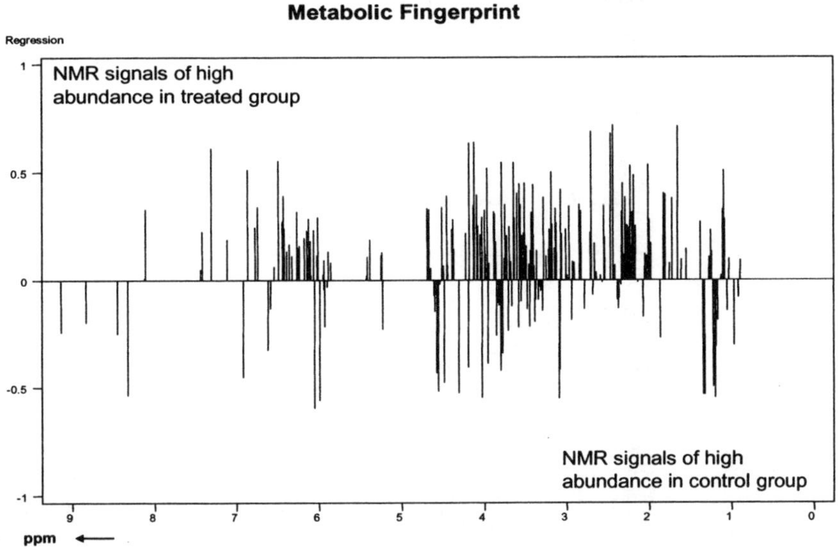

Figure 3. *A factor spectrum which is used to identify the metabolite NMR peaks that differ in rat urine upon bromobenzene treatment compared to controls.*

Data processing and bioinformatics - The data from the large-scale experiments require powerful data processing equipment and algorithms. Data processing methods like standardization, normalization or scaling are applied in order to be able to compare the obtained measurements. Systematic biases originating from various technological sources are corrected, and the focus is directed towards the differences in the measurements are that are determined by the biological parameters under study. A new challenge for biologists is to turn large datasets with high amounts of noise and without obvious biological meaning into relevant findings, that includes selection of subsets of the genes, proteins or metabolites from the thousands, that have biologically relevant characteristics, and to further investigate only these subsets. Techniques are applied for clustering of the data into interpretable groups or structured patterns. Methods applied for this purpose include various clustering algorithms that calculate a measure of similarity between the expression profiles of the genes. Methods include hierarchical clustering, as shown in figure 4, K-means clustering, self organizing maps (SOM) that are applied to create clusters of genes that behave more similar to each other than to genes outside of the cluster. The number of clusters formed can be imposed upon the dataset or can be determined by the clustering algorithms automatically. Once smaller subsets have been found that account for most of the changes introduced by the environmental conditions, the molecules in these subsets are further analyzed with respect to biological relevance. Reviews on different methods of microarray data analysis: [8], [9]

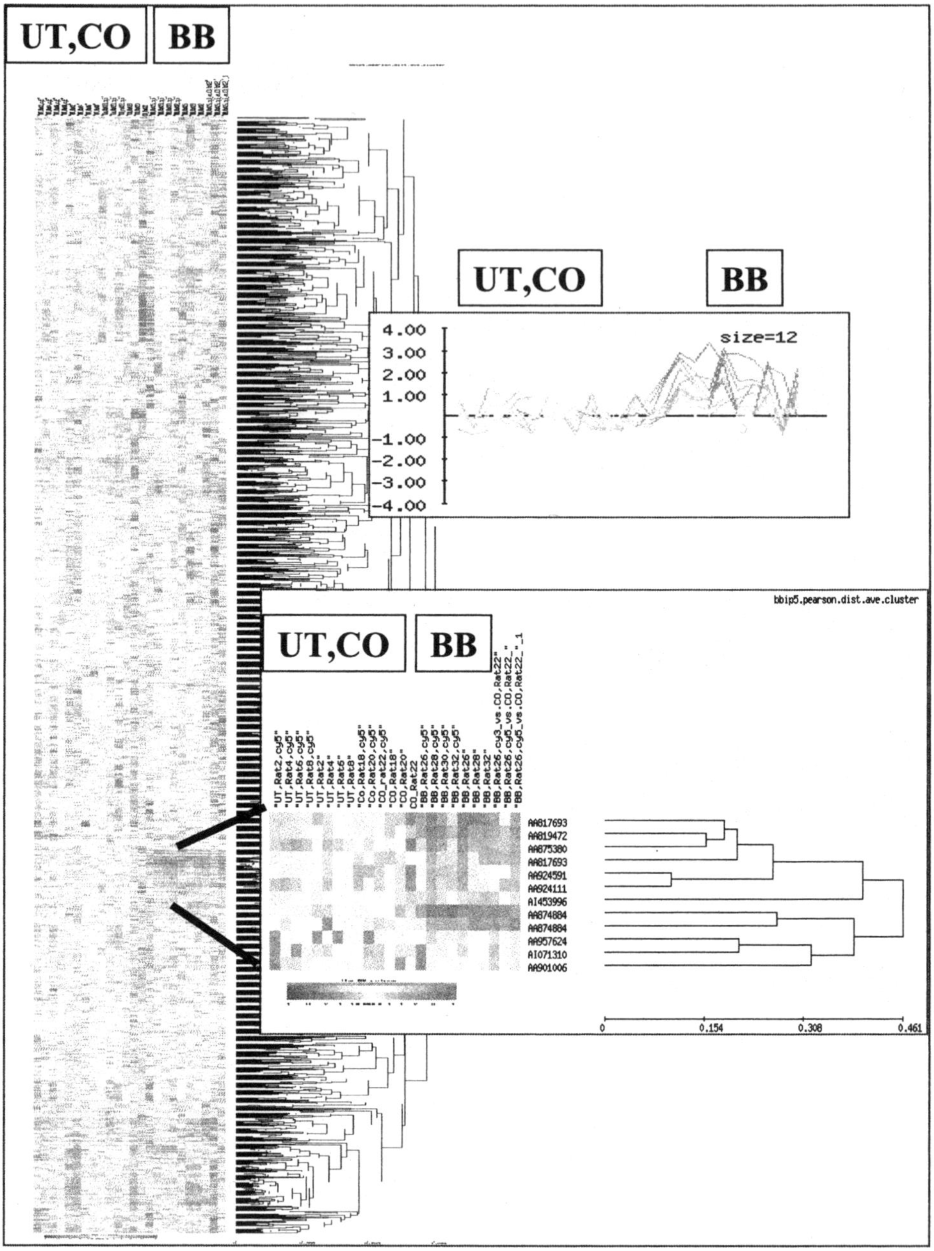

Figure 4. Dendrogram of hierarchically clustered transcriptomics data of rat liver of bromobenzene (BB) treated rats and untreated (UT) and corn oil (CO) controls. Columns represent the individual samples from the different groups, while the rows represent the genes. One specific cluster of genes upregulated by BB is enlarged.

Unsupervised methods such as principal component analysis (PCA) determine intrinsic structure within data sets, without prior knowledge. With such methods, a direct comparison of datasets, for instance NMR spectra, is made and subsets of data are formed, solely on the basis of similarities of NMR spectra. Supervised methods such as partial least squares (PLS) and principal component discriminant analysis (PCDA) are more powerful tools, which use additional information on the data set such as biochemical, histopathological or clinical data to identify differences between pre-defined groups.

Biological interpretation- As mentioned, it is not feasible to analyze the expression data one-by-one, on a single gene or protein basis. Moreover, doing so would result in a great loss of the information which resides in the coherence of the data collected in one study. The relationships between genes or proteins expressed in a certain situation regarding time, localization and experimental conditions are the most valuable information obtainable in functional genomics studies. Interaction of proteins with other proteins, ligands, cellular metabolites, DNA, or RNA. Studying the interaction and integration of the individual components is of crucial importance for the understanding of biology.

1.3 Functional genomics in toxicology

Interindividual genetic differences (e.g., SNPs in specific genes) can be of great importance in toxicology. If a drug-metabolizing enzyme is affected by the genetic abnormality, the catabolizing activity of the enzyme could be altered. The rate of compound activation and metabolism and the mechanisms of protection determine to what extent toxicity is found in an organism. Genetic variations may be of great importance in all of these processes. The genetic differences between human and species that are used to predict potential toxicity (in human) are to be characterized in order to better extrapolate results from toxicity testing to the human situation. The determination of thousands of gene or protein expression levels simultaneously in a given sample provides an insight in the molecular processes that together determine the specific status of that sample. Toxicity can be seen as the distortion of biological processes in an organism, organ or cell. Thus, by investigating the cellular-molecular mechanisms in a cell, functional genomics technologies can be of extreme value in toxicology as in other life sciences. Gene expression profiles can be used to discriminate samples exposed to different classes of toxicants, to predict toxicity of (yet) unknown compounds and to study cellular mechanisms that lead to or result from toxicity. [10], [11], {72} Indeed, transcriptomics was shown to be powerful in the mechanistic assessment of toxic responses [12]; [13;14]; [15;16]

Toxicity fingerprinting and prediction- Detection of toxicity at an earlier stage, in the screening of many drug candidates may be a very advantageous application of functional genomics technologies. Pharmaceutical companies are developing new candidate active compounds at extremely high rate, for instance using technologies like combinatorial chemistry. Most of the thousands of newly synthesized compounds will never reach the market, as only a few will be selected as potential drugs and those are evaluated in a time and cost-consuming process looking at both efficacy and toxicity. High throughput screening of drug candidates for potential signs of toxicity may provide a useful criterion for early selection. The measurement of gene or protein expression profiles upon compound exposure can, like a fingerprint, be used to classify the compound according to similarity in profile to exposure upon known (model) toxicants. Large databases of these expression profiles enable the classification of the expression pattern from the sample of toxicological interest according to e.g., the toxic potency, type of mechanism, target organ, dose and time of exposure. A new compound can thus be identified as putatively toxic based on the common mechanisms of response at the molecular level. Besides this (semi-) high throughput screening for (any) toxic properties of new compounds, functional genomics technologies could be used to quickly identify toxicity target organs or tissues without prior knowledge. The alteration in expression of genes specific to certain tissues can be used as an indicator of the involvement of this tissue in toxicity responses.

Markers of toxicity - The finding of new specific and indicative marker genes or proteins that can detect or even predict certain types of toxicity is a major progress effected by transcriptomics and proteomics. However, if toxicologists want to be able to discern different mechanisms that eventually may lead to the same symptoms of toxicity, single gene or protein markers will not be sufficient. More likely, subtly altered expression levels of many genes together can define the status of a cell, thereby requiring precise and large-scale measurements of the pattern of expression of thousands of genes or proteins. A cell-wide pattern of gene or protein expression, in analogy to a fingerprint, can be used to discern a healthy cell from the different stages of distortion from this normal status.

Mechanism elucidation - Interactions of genes and proteins underly the majority of the biological processes, and coordinate expression of genes or proteins under specific circumstances provides an indication of a certain relationship in a biological mechanism. The molecular mechanism of toxicity can thus be reconstructed through the observed changes in genes that interact with or influence each other. As thousands of genes are investigated simultaneously, the chance of finding the target molecule that triggers a toxic effect upon interaction with the xenobiotic is greatly improved. This target molecule for instance can be receptors that initiates an adverse effect in the cell upon binding of the (xenobiotic) ligand. Moreover, transcriptomics experiments can be designed to identify the organs or organelles in an organism that most likely form the target sites of adverse effects. After an initial organism-wide assessment of effects, characteristic changes in gene expression can indicate which organs should be chosen for further toxicological examinations.

Interspecies comparisons - Even if laboratory animals and humans are very different in many aspects, there is a high similarity at the molecular level. A majority of the genes found in humans is also present in other organisms like rat or mouse. Some genes are identical between different species, and overall, genomes of man and rodents exhibit more than 90% similarity. Obviously, this is also the reason why these animals are frequently used as model-substitutes for humans in experimental settings. In functional genomics, the similarity at the molecular level will be of even more benefit for the extrapolation of results between species. Although the physiological responses might still be different, the underlying molecular mechanisms may show a much higher degree of conservation between species. The same argument applies for extrapolation from in vitro experiments to the *in vivo* situation. Even though the physiological effects will not be identified *in vitro*, the underlying molecular mechanisms eventually leading to these effects might be very similar *in vivo*. Moreover, the circumstances for performing experiments *in vitro* can be controlled and monitored in a more precise manner, enabling to focus on the effects of changes in the conditlions of interest only. As always, *in vitro* experimentation requires a thorough validation of the biological significance of the findings in the human situation.

Mixture toxicology - The assessment of the harmful effects caused by substances in complex mixtures, such as environmental pollutants, is currently hardly feasible. In order to assess combinatorial effects of compounds, large studies have to be designed including both individual exposures as well as many combinations. Since only a few markers of toxicity are monitored, discrimination between effects in the different groups is limited. As functional genomics technologies allow the monitoring of thousands of effects, the likelihood of finding different effects between exposures is greatly increased. Moreover, mechanisms of the combinatorial effects can be studied and/or predicted at the molecular levels. Synergistic or additive effects can be expected when similar molecules or pathways prove to be the targets of different substances in the mixture. Similarly, when changes occur in different, specific biological pathways, this could indicate that interference should not be expected from the compounds in the mixture.

Metabonomics in toxicology - Measurement of metabolite levels can serve multiple goals in toxicology: One is to identify (and quantify) breakdown products of the toxic compound after metabolism in the organism. The clearance of the toxicant can be determined from the excreted metabolites in the urine. The recovery of (metabolites of) the toxic compound in the urine can be used to deduce levels of exposure and confirm successful dosing of the animals. The population of endogenous metabolites in a cell can be used as a fingerprint of the (health) status of a cell. Metabolites can be identified that may serve as markers of disease or toxicity. Levels of (combinations of) metabolites can be applied to accurately discern between different health or toxicity states.

1. 4 Molecular mechanisms in hepatotoxicity

Processes leading to, and resulting from toxicity of the liver are currently monitored on a routine basis using a wide variety of mostly empirically determined parameters. Changes in the liver relative weight and gross pathological observations like color or texture of the tissue are rough indicators for toxicity. Specific liver toxicity endpoints include necrosis, hypertrophy, hepatitis, cholestasis, hyperplasia and steatosis. All of these involve more or less specific changes at the cellular molecular level. In the process of necrosis, liver cells are ruptured and membrane damage leads to leakage of normally intracellular enzymes to the extracellular fluids. The presence in blood plasma of those enzymes can be determined using a specific enzyme activity assay as a quantitative indicator of the extent of damage. Enzymes like alkaline phosphatase (ALP), lactate dehydrogenase (LDH), and aminotransferases (ALAS and ALAT) are commonly used as indicators of liver necrosis. Along with the leakage of enzymes from necrotic cells, blood plasma levels of abundant cellular proteins, like serum albumin and creatin have been related with disruptions of homeostasis. Besides enzymes and proteins, small biomolecules (metabolites) metabolized in cellular pathways such as gluconeogenesis, fatty acid metabolism, amino acid metabolism can be indicative of perturbed cellular processes. Most of the widely measured clinical chemistry parameters were established empirically. Rather than mechanisms of toxicity, they are often the result of aberrant cellular processes. While leakage of enzymes through disrupted membranes clearly is a secondary process in a late stage of cell death, other parameters like protein levels might be directly changed as a result of alterations in the cellular mechanism in which these proteins play a role. Those parameters are very likely to be reflections of changes in the expression levels of genes or proteins that form the basis of the cellular response towards a change in the environment of the cell. Large-scale determination of gene or protein expression levels will enable the embedding of currently used observations in a cell-wide pattern of changes.

2. Toxicogenomics of hepatotoxicity: case studies

To illustrate the use of functional genomics technologies in toxicology, two studies are described that aim to investigate molecular mechanisms of hepatotoxicity. The first deals with acute (24 hours) liver toxicity induced by bromobenzene, which has served as a model compound for toxicologists for a long time now. Another study performed in our laboratory aimed at the detection of the effects of a low-level exposure to mixtures of toxicants. Rats were exposed for 28 days to low levels of food additives which were found to cause adverse hepatotoxic effects at higher concentrations (evaluated in [17]). The effects on the liver induced by the food additives are expected to be subtle and the ability to observe changes at the level of gene expression are explored. The results of the transcriptomics experiments for four individual compounds are to be used for inference of (mechanistic) interactions upon combinatorial exposure to these food additives. This combined intake of mildly toxic food additives might pose a realistic health risk to humans.

2.1 Acute hepatotoxicity induced by bromobenzene

The well-studied toxicant bromobenzene (BB) was used to induce hepatotoxicity in rats [18]. Bromobenzene is an industrial solvent that elicits toxicity predominantly in the liver, where it causes centrilobular necrosis. As bromobenzene is a hydrophobic molecule, it requires biotransformation to increase water-solubility for excretion in the urine. Bromobenzene metabolism and toxicity in (rat) liver have been described in detail. [19],[20],[21],[22],[23] Biotransformation includes a cytochrome P450-mediated epoxidation, and the epoxide metabolites are either conjugated to glutathione, hydrolyzed enzymatically by epoxide hydrolase or further oxidized, leading to hydroquinone-quinone redox cycling. At high BB doses, primarily due to conjugation to the epoxides or bromoquinones, hepatic cellular glutathione is depleted. If [GSH] falls below a threshold level, cell damage occurs from the non-sequestered metabolites. Bromobenzene toxicity has been related to the covalent binding of reactive BB metabolites to endogenous proteins, especially containing sulfhydryl groups [24]. GSH depletion elicits a number of secondary reactions like lipid peroxidation ultimately leading to cell death.

Rats received an intraperitoneal dose of bromobenzene, which was chosen to be highly hepatotoxic, as confirmed by the finding of nearly complete glutathione depletion at 24 hours after bromobenzene administration (figure 5B). The low level of oxidized (GSSG) relative to reduced glutathione (GSH) indicate that the depletion is due to rather than to oxidation of glutathione. The bromobenzene administration resulted in an average decrease in body weight of 7 % after 24 hours, whereas vehicle control rats gained on average 6 % of weight (figure 5A).

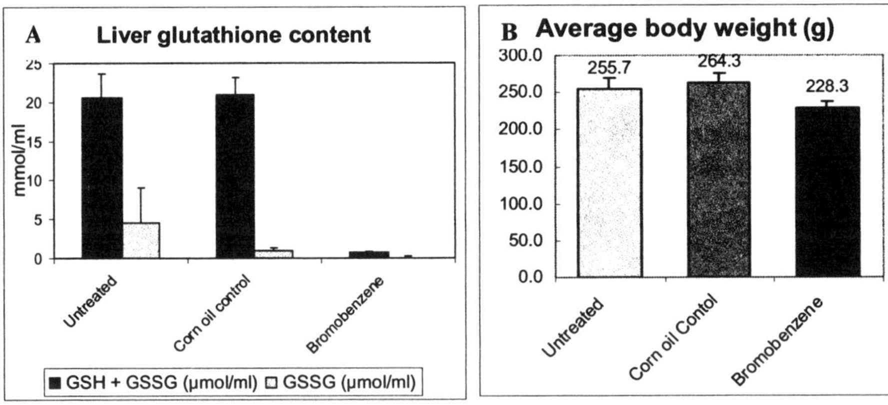

Figure 5. A. Body weight changes and **B.** Glutathione (GSH) depletion in liver of rats treated with brombenzene, corn oil as the vehicle control, or untreated controls.

Transcriptomics - cDNA microarray measurements were routinely performed for every sample in duplicate, measuring mRNA levels of about 3000 different genes. The majority of the genes represented on our microarray was comparably expressed throughout all the samples. However, the bromobenzene treatment distinctly elicited alterations in the expression pattern of a number of genes in rat liver. Principal component analysis (PCA) (Figure 6) visualizes the differences in the expression profiles of rat liver after BB treatment compared to the controls, while the corn oil injection controls could not be distinguished from the untreated controls. The interindividual biological variation within the treatment groups did not exceed the technical variation. The genes that account for most of the differences in the expression profiles of BB-treated compared to controls in the PCA were found to be the genes that were also identified by fold-change calculations. Average fold changes were calculated after BB treatment compared to controls, and statistical significance of the changes was determined using Student's T-test where the threshold for significance was a p-value smaller than 0.01. The genes that changed more than 1.5 fold (up or down) upon BB treatment (with $p<0.01$) are listed in table 1 (upregulated) and 2 (downregulated). The importance of the depletion of intracellular GSH levels in bromobenzene hepatotoxicity was shown the gene expression level with the finding of the differential expression of genes involved in GSH metabolism. The GSH-S-transferase Alpha, Mu and Pi subunits, catalyzing the conjugation of reactive metabolites to GSH, were all upregulated. The induction of γ-glutamylcysteine synthetase, which is the rate-limiting enzyme in GSH synthesism suggests GSH synthesis. The other enzyme important in the this process, GSH synthase, was not found to change significantly at the mRNA level, but at the protein level, a bromobenzene-induced increase in GSH synthase was found. A significant average 1.4-fold increase in activity of glutathione-S-transferase towards the substrate CDNB was observed in the cytosolic liver fractions of rats treated with bromobenzene compared to controls. The induction of microsomal epoxide hydrolase is coherent with its role in bromobenzene metabolism; the hydrolysis of epoxide BB intermediates. It was also found to be induced by xenobiotics like phenobarbital, trans-stilbene oxide and Aroclor 1254 [25]. Other enzymes involved in drug-metabolism like the aflatoxin B1 aldehyde reductase, aldehyde dehydrogenase and the NADPH-cytochrome P-450 oxidoreductase were significantly upregulated. The co-ordinate response of a cluster of genes strongly upregulated by BB (GSTA, GSTM, γ-GCS, HO-1 and ferritin, peroxiredoxin1 and NADPH-cytochrome P450 oxidoreductase) suggests a common regulatory mechanism. Indeed, many of the genes in this cluster have been shown to be under transcriptional control of the Electrophile Response Element (EpRE, formerly named antioxidant response element, ARE) [26], [27]. Genes typically responding to oxidative stress such as heme oxygenase-1, peroxiredoxin1 metallothioneins, ferritin, TIMP1 were found to be heavily induced. HO-1 is the enzyme that catalyses the degradation of heme into biliverdin, and induction could be functional against oxidative stress. [28], [29]. Bilirubin, produced from biliverdin by biliverdin reductase is an antioxidant, which would provide the cell with a defense mechanism against reactive oxygen species. The intraperitoneal administration of a high dose of bromobenzene clearly elicits a response well known as the acute phase response. Especially in the liver, the APR coincides with dramatic changes in a wide variety of proteins with the general purpose to provide a protective response and re-establish cellular homeostasis. [30]

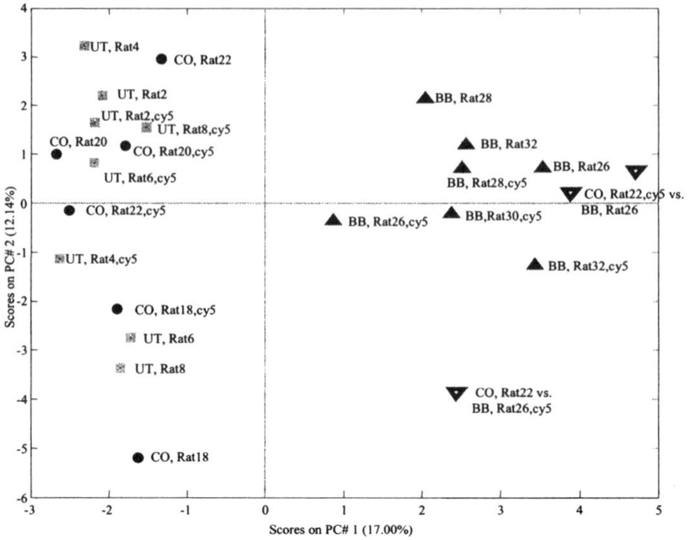

Figure 6. Principal component analysis (PCA) visualizes the differences in the expression profiles of rat liver after BB treatment compared to the controls, while the corn oil injection controls could not be distinguished from the untreated controls.

Table 1: Liver genes significantly (p<0.01) upregulated by bromobenzene

AC#	Gene name	Average CO / UT		Average BB / CO	
		Fold Change	P-value	Fold Change	P-value
AA900551	epoxide hydrolase (microsomal EH 1)	0.76	0.340	5.59	0.000
AA874884	heme oxygenase gene	0.99	0.976	5.20	0.001
AA923966	AFAR aflatoxin B1 aldehyde reductase	1.04	0.766	4.70	0.000
AA817693	ferritin light chain subunit,	0.83	0.391	4.29	0.005
AI059976	Vesl-2	0.94	0.739	3.77	0.001
AA818339	GSTA class alpha ; liver glutathione S-transferase Ya subunit, clones pGTR1	0.84	0.101	3.70	0.000
AA964496	Highly similar to actin beta	1.14	0.585	3.48	0.005
NA	cathepsin L	0.94	0.622	3.21	0.007
AA998895	nucleolar protein B23	0.95	0.467	3.02	0.000
AA858998	Highly similar to Chain C, Gdpran-Ntf2 Complex ; RAN, member RAS oncoge	0.91	0.740	2.88	0.004
AA900218	metallothionein-i (mt-1)	1.46	0.006	2.76	0.002
AA859399	Metallothionein	1.86	0.008	2.71	0.002
AA874884	heme oxygenase gene, HOX	1.06	0.699	2.70	0.002
AA875245	Peroxiredoxin 1 (Heme-binding protein 23)	0.96	0.805	2.61	0.000
AA998734	GSTM soluble,class Mu 1 (2?) ; glutathione S-transferase (GST) Y(b) (EC 2.5	0.97	0.923	2.58	0.004
AA965220	GCSI ; gamma-glutamylcysteine synthetase light subunit	1.07	0.625	2.53	0.006
AI045792	Highly similar to 60S RIBOSOMAL PROTEIN L9	1.14	0.537	2.47	0.010
AI030685	nestin	0.98	0.917	2.44	0.007
AA818339	GSTA soluble,class Alpha ; liver glutathione S-transferase Ya subunit, clones	0.94	0.509	2.40	0.006
AI060085	thyroid stimulating hormone receptor ;Thyrotropin receptor	1.10	0.426	2.36	0.002
AA964725	Weakly similar to pervin ; Highly similar to human and mouse Flightless-I prot	1.18	0.615	2.35	0.008
AA819544	ribosomal protein S2	1.04	0.802	2.33	0.001
AA819244	nuclear-encoded mitochondrial elongation factor G	0.51	0.051	2.31	0.004
AA859484	Highly similar to PROTEASOME THETA CHAIN	0.96	0.654	2.27	0.000
AA900657	60S RIBOSOMAL PROTEIN L7A	0.96	0.664	2.25	0.000
AA965078	mitochondrial 3-2 trans-enoyl-CoA isomerase	1.01	0.876	2.22	0.006
AA955106	aldehyde dehydrogenase (ALDH)	1.09	0.784	2.20	0.005
AI045953	cysteine sulfinic acid decarboxylase	1.02	0.949	2.17	0.004
AA858741	apolipoprotein M ; Highly similar to NG20 [M.musculus]	1.21	0.253	2.05	0.001
AA875245	Peroxiredoxin1 (Heme-binding protein 23)	0.92	0.413	2.04	0.001
AI059951	neuronal cell death related gene in neuron -7 (DN-7)	1.10	0.124	2.01	0.001
AA818441	ferritin-H subunit	0.89	0.415	2.01	0.004
AA874997	ribosomal protein S8	1.04	0.791	1.97	0.001
AA818887	MHC class I	0.92	0.509	1.96	0.004
AA874949	Highly similar to INITIATION FACTOR 5A [H.sapiens]	0.88	0.448	1.95	0.001
AA924912	ribosomal protein L6	0.86	0.395	1.94	0.000
AA956323	membrane bound cytochrome b5	0.78	0.225	1.90	0.000
AI058620	(tc2) for TEGT; Bax inhibitor-1 (BI-1) (Testis enhanced gene transcript)	1.09	0.627	1.89	0.002
AA899448	ribosomal protein S3a	0.78	0.448	1.89	0.006
AA818640	peptidylarginine deiminase mRNA / Ribosomal prot 40S	0.88	0.451	1.88	0.006
AA925200	60S ribosomal subunit protein L35	0.93	0.588	1.85	0.007
AI058566	peptide/histidine transporter	1.06	0.669	1.82	0.001
AA874955	clathrin light chain (LCB3)	0.78	0.063	1.80	0.001
AA900595	Highly similar to CGI-98 protein [H.sapiens] anti-silencing protein ASF1 homo	1.10	0.241	1.80	0.000
AA964548	Highly similar to TUBULIN BETA CHAIN	0.99	0.949	1.79	0.000
AA956507	NADPH-cytochrome P-450 oxidoreductase	0.77	0.268	1.78	0.002
AA866442	somatic cytochrome c	0.64	0.227	1.77	0.006
AA900102	Mitochondrial import inner membrane translocase subunit TIM8 A (Deafness	1.00	0.982	1.74	0.005
AA899102	glucose-6-phosphate dehydrogenase (Gd, EC 1.1.1.49)	0.98	0.861	1.73	0.003
AA874837	phosphoglycerate mutase B isozyme (PGAM)	0.82	0.195	1.73	0.007
AA819314	Highly similar to METHIONINE AMINOPEPTIDASE 2	0.78	0.129	1.72	0.010
AA866268	Highly similar to RL8_HUMAN 60S RIBOSOMAL PROTEIN L8	1.10	0.391	1.71	0.002
AA818442	ribosomal protein S11	0.89	0.174	1.70	0.001
AA859476	cofilin	0.81	0.166	1.69	0.006
AA957593	Tissue inhibitor of metalloproteinase-1 (TIMP1)	0.97	0.772	1.69	0.004
AI453996	Heat shock 70kD protein 5 / GRP Glucose regulated protein	0.96	0.720	1.68	0.004
AA819204	Highly similar to METALLOTHIONEIN-IV (mouse); weakly similar to METALLO	1.14	0.446	1.66	0.007
AA964657	EST, unknown	0.96	0.744	1.64	0.005
AA875102	Highly similar to U1 AND U2 SMALL NUCLEAR RIBONUCLEOPROTEIN E [M	0.85	0.263	1.62	0.001
AA899472	Highly similar to translation intiation factor eIF-3 p110 subunit [H.sapiens]	0.89	0.410	1.58	0.000
AA875070	alpha-prothymosin	0.94	0.659	1.58	0.006
AI043642	potassium channel (KCNQ2)	1.00	0.978	1.53	0.003
AA900527	ribosomal protein S26	0.92	0.451	1.52	0.009

Table 2: Liver genes significantly (p<0.01) downregulated by bromobenzene

AC#	Gene name	Average CO / UT Fold Change	P-value	Average BB / CO Fold Change	P-value
AA901050	growth hormone-regulated protein SPI 2.3 (GHR-P63) (hepatocyte growth hor	0.66	0.201	0.32	0.000
AA964788	GPX1 ; glutathione peroxidase	1.10	0.554	0.37	0.001
AA819810	GST ?; Weakly similar to GLUTATHIONE S-TRANSFERASE P	1.14	0.202	0.38	0.000
AA963445	selenoprotein P	1.10	0.619	0.39	0.002
AA926359	receptor-linked protein tyrosine phosphatase (PTP-PS)	1.18	0.305	0.39	0.000
AA819164	mitochondrial IF1 ; ATPase inhibitor	0.96	0.848	0.41	0.005
AA858662	14-3-3 zeta isoform ;Tyrosine 3-monooxygenase / tryptophan 5-monooxygena	0.82	0.175	0.41	0.000
AA924288	thymosin beta-10 gene	0.77	0.399	0.42	0.005
AI072074	Fatty acid binding protein 1, liver	1.09	0.687	0.42	0.006
AA819465	apolipoprotein A-I gene / preapolipoprotein CIII ?	0.96	0.803	0.44	0.000
AA901407	betaine homocysteine methyltransferase (BHMT)	0.67	0.057	0.45	0.003
AA926010	long-chain acyl-CoA synthetase (EC 6.2.1.3)	1.12	0.543	0.46	0.002
AA858975	transferrin	0.92	0.616	0.46	0.006
AI137907	alpha-1-macroglobulin	1.22	0.340	0.46	0.006
AA819810	GST ?; Weakly similar to GLUTATHIONE S-TRANSFERASE P	1.32	0.063	0.46	0.001
AA819043	Fibronectin 1	0.88	0.389	0.47	0.000
AA963164	synaptojanin II	1.00	0.987	0.49	0.004
AI044677	Weakly similar to gamma-fibrinogen	1.84	0.010	0.51	0.004
AA819267	fibrinogen gamma chain-a	0.98	0.888	0.52	0.007
AA901379	plasma protein; inter-alpha-inhibitor H4P heavy chain	1.52	0.045	0.52	0.009
AA819784	Hemoglobin, alpha 1	1.39	0.165	0.52	0.010
AI136048	mitochondrial 3-hydroxy-3-methylglutaryl-CoA synthase	1.45	0.094	0.53	0.008
AA900592	Weakly similar to serine protease; similar to complement component Clr, Mm	1.23	0.136	0.53	0.000
AA956238	vitronectin	1.29	0.154	0.54	0.002
AA819034	ESTsWeakly similar to AF154572_1 ERG2 protein ; alpha interferon inducible	1.31	0.155	0.54	0.008
AA998796	complement component C9 precursor	1.21	0.102	0.54	0.001
AA963258	insulin-like growth factor I (IGF-I)	1.12	0.528	0.55	0.009
AA817963	negative acute-phase protein alpha-1-inhibitor 3 mRNA, 3' end	0.66	0.049	0.55	0.005
AI072074	Fatty acid binding protein 1, liver	0.85	0.432	0.56	0.001
AA818579	cysteine dioxygenase	0.76	0.175	0.58	0.009
AA925346	CYP51; lanosterol 14-alpha-demethylase	1.10	0.257	0.61	0.000
AA817963	negative acute-phase protein alpha-1-inhibitor 3 mRNA, 3' end	0.62	0.007	0.64	0.008
AI070597	Highly similar to CGI-97 protein [H.sapiens]	1.60	0.002	0.65	0.009
AA866264	Weakly similar to 20-ALPHA-HYDROXYSTEROID DEHYDROGENASE	0.92	0.524	0.66	0.010
AA957012	Highly similar to Human RNA polymerase II chain	1.00	0.980	0.68	0.001

Proteomics - Livers of the rats subjected to an intraperitoneal injection of bromobenzene were also used to measure protein profiles. Proteomics technology using two-dimensional gel electrophoresis was used to separate the cellular protein samples. After separation, image analysis software was used to quantify the intensity of the spots on the gel-photographs. Three 2-dimensional gels were prepared from each liver extract. In Figure 2A, the protein pattern obtained from an untreated animal is shown. Spot numbers of proteins that were differentially expressed in bromobenzene or corn oil treated animals are indicated. Figure 2B shows a typical 2-dimensional gel of liver proteins of a bromobenzene-treated rat. Figure 7 displays the volumes for all spots identified as statistically significantly changed upon bromobenzene or corn oil treatment. The bars represent normalized spot volumes, averaged for all (9-12) gels within one treatment group (untreated, vehicle control or bromobenzene treated). Hence, the dispicted standard deviation reflects both the interanimal variation, as well as technical variation between gels. Spots marked as significantly changing were identified for analysis by MALDI-TOF or nano-electrospray mass spectrometry. At the protein level, bromobenzene induced a 2.5-fold increase in the mitochondrial aldehyde dehydrogenase 2. A decrease in Heat shock protein 60 (HSP60) and ATPase beta subunit protein was found. It was recently shown that also acetaminophen-induced hepatotoxicity resulted in decreased ATP-synthase subunit and HSP60 levels in mice [31]. The amount of 4-hydroxyphenylpyruvic acid dioxygenase (HPD) and D-dopachrome tautomerase decreased upon BB treatment. Remarkably, two spots represent HPD proteins with similar molecular mass but different pI, suggesting that HPD exists in more than one state, possibly through post-translational protein modifications. The decrease in the carbamoyl-phosphate synthase (CPS) of ~40 kDa and the concordant increase in the smaller (~25 kDa) CPS protein upon BB treatment propose a degrading activity towards this protein and provide an example of the capability to visualize protein processing with 2-dimensional gel electrophoresis. The proteomics data (figure 2A and B) show that bromobenzene specifically induced a shift in the protein pattern from high molecular mass proteins towards more lower molecular mass, which could point out a bromobenzene-induced degradation of proteins. Corroborating, the mRNA levels for the lysosomal proteolytic enzyme cathepsin L significantly, as well as many subunits of the proteasome complex increased upon BB injection. This, in combination with the induced transcription of many factors involved in protein synthesis such as ribosomal proteins, translation initiation and elongation factors, suggests a higher protein turnover rate. Principal component analysis on the proteomics data, as for the transcriptomics data, illustrated a clear difference between liver protein patterns of bromobenzene-treated animals and controls.

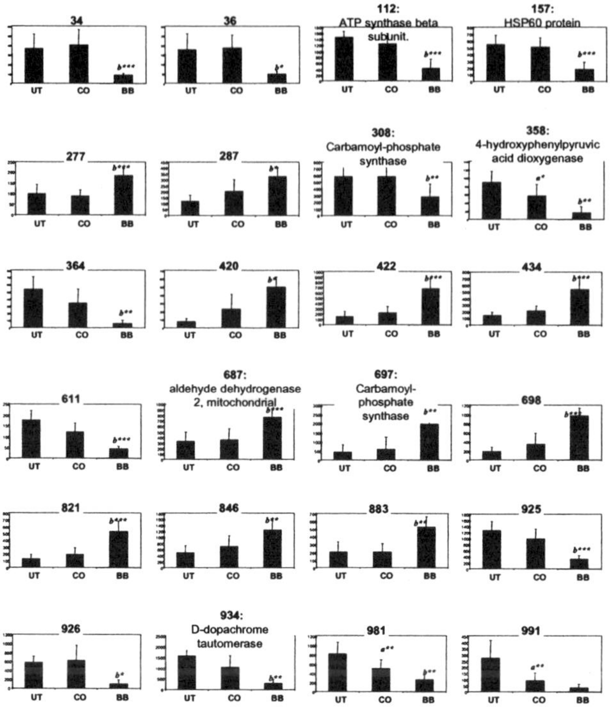

Figure 7. Volumes for all spots identified as statistically significantly changed upon bromobenzene or corn oil treatment. The bars represent normalized spot volumes, averaged for all (9-12) gels within one treatment group (untreated, vehicle control or bromobenzene treated).

Metabonomics - Urine and blood plasma were collected from bromobenzene-exposed rats. Samples were measured in triplicate using NMR, and the NMR spectra were processed using a partial linear fit algorithm. NMR spectra of both urine and plasma from bromobenzene-treated rats were distinct from control samples. Mathematical pattern recognition techniques have been developed in order to quantify the dissimilarity between two patterns. The principal components were calculated for each NMR-spectrum obtained from urine of the rats. The first two principal components could be used to clearly distinguish the spectra of the bromobenzene treated from the controls. Besides calculating the difference between two spectra, mathematical techniques can be applied to locate which signals in the spectra are mainly responsible for the pattern differences. Elaborate studies are required to identify the positions of the peaks of the many bromobenzene-metabolites in the spectra. This will enable the further elucidation of the mechanisms of bromobenzene-biotransformation and excretion. Literature reports describe diverse pathways in the biotransformation of bromobenzene to bromphenols, bromoquinones and glutathione-conjugated bromobenzene intermediates. The quantification of specific metabolites in liver, plasma or urine can be used to deduce whether roles of the enzymes, such as different CYP species, GSTs and mEH in the biotransformation pathways.

2.2 Integration of functional genomics experimental results

The simultaneous measurement of the expression of thousands of genes and proteins in liver, along with the metabolite profiles in urine and plasma yield a tremendous amount of information with respect to hepatotoxicity in rat upon BB administration. The integration of the datasets poses the next challenge in functional genomics research. Whereas with transcriptomics experiments, the genes under study are known on beforehand, proteomics and metabonomics experiments result in the finding of thousands of proteins or metabolites (NMR signals) that have to be identified for a biological interpretations are possible. The labor-intensive proteomics techniques using 2D-gel based separation and mass spectrometry enable the identification of only limited numbers of proteins. The identification is also the rate-limiting step for metabonomics experiments. Therefore, direct matching of the results with transcriptomics data is severely hampered. Using proteomics methods, only 24 proteins were found to statistically significantly change in all animals upon bromobenzene treatment, which is in contradiction to the massive shift in the proteome pattern observed in the gel images and the number of genes with differential expression upon bromobenzen treatment. This restricted number can mainly be ascribed to the strict quality criteria imposed on the data before genes or proteins were called as differentially expressed after treatment. Moreover, technical limitations of the 2-dimensional gel electrophoresis in terms of protein abundance, solubility and physico-chemical properties prevented the detection of many proteins for which corresponding mRNA level changes were observed using transcriptomics. Furthermore, for about half of the proteins that were found to significantly change, the identification by mass spectrometry was not successful. Whereas protein degradation was concluded from the 2D-gel analysis, the upregulation of gene expression of components of the proteasome, cathepsin and ribosomal subunit point towards increased protein degradation or turnover. Individual metabolites were not (yet) identified, while the entire patterns by itself proved to be highly useful in identification of toxicity. Especially in the urine, the presence of putatively BB-related degradation products (metabolites) was obvious and strongly influencing the total metabolite profiles, depending on the administered dose and time of sample collection. Further work attempts to identify specific metabolites that could be useful as markers of both the exposure and of hepatotoxicity. In general, multivariate pattern recognition techniques prove to be essential in the analysis and interpretation of large datasets obtained through the various functional genomics technologies in toxicology.

2.3 Hepatotoxicity induced by food additives

A study was recently initiated in our laboratory in order to show the value of functional genomics techniques in the assessment of the molecular effects of mixtures of chemicals. In order to be able to describe the effects of food additives on the gene expression in the liver, transcriptomics experiments are performed on individual food additives. Subsequently, the effects of mixtures will be studied. Intake levels of individual food additives are considered to be safe at no observed adverse effect levels (NOAEL). However, the current NOAEL used for food additives is most often based on empirical sciences, e.g., pathology, clinical chemistry and limited in vitro studies. The outcome of these studies is usually only descriptive (e.g., cytotoxicity, LD50, genotoxicity) and poorly comprehensive from a mechanistic point of view. Even if mechanistic information is available for the action of food additives, this is limited to the description of the effects of the compound on a limited number of genes (e.g., hepatic phase I and phase II metabolizing enzymes). Little is known about potential health effects under those circumstances where additives are taken simultaneously at or below the NOAEL. As described. [17], possible joint actions in the liver could not be excluded for four food additives. Moreover, extrapolation of the results from animals to humans, as well as the risk assessment using these results up to date has been performed on a theoretical basis. This makes extrapolation of the results to the human situation uncertain and relies on imposed safety factors. Insights in the molecular mechanisms of the toxicity obtained through transcriptomics measurements might provide a more profound basis for assumptions when extrapolating data from animal experiments to the human situation. Using transcriptomics, an attempt is made to evaluate hazards which are possibly introduced by interaction of the food additives. Rats were dosed with the 4 different food additives at three four-dose levels around the reported NOAEL. The samples chosen for transcript profiling were exposed to concentrations below the doses where obvious toxicity was observed, e.g., liver weight differences or histopathologically identified aberrancies. Pooled RNA samples of rat livers were monitored for the expression of about 3000 genes using cDNA microarrays. Preliminary data indicate that the expression of phase I drug metabolizing enzymes was induced upon treatment with the food additives, in good agreement with present knowledge. In addition, changes in mRNA level of these drug-metabolizing enzymes were confirmed by means of RT-PCR, providing a technical validation of the mRNA expression changes. Food additive induced changes in the expression of phase I drug metabolizing enzymes were also successfully validated at the protein and functional level by means of western blotting and specific enzyme activity assays. These data indicate that changes in hepatic transcription can be predictive for actual physiological changes that may occur when rats are exposed to food additives. Besides phase I drug metabolizing enzymes, many other genes with implications in various biological processes were identified to be altered in the liver upon exposure to (some of) the compounds. Gene expression changes already occurred at concentrations were no signs of hepatotoxicity were observed. Taken together, these studies show the value of transcriptomics in terms of predicting changes that occur at physiological level. In a broader sense, these data demonstrated that a transcriptomics approach allows predictions of mechanism of hepatotoxicity of food additives and of possible interactions.

3. Conclusions

Case studies in toxicology were presented which show that using transcriptomics, proteomics and metabonomics, a variety of new findings is obtained even when using a well-known model compound like bromobenzene. Conclusions on molecular mechanisms of action are deduced from the data, and effects can be determined at lower doses compared to conventional toxicity parameters. It is of crucial importance that the findings from functional genomics experiments are integrated in the "classical" toxicology. Described findings and published results from other laboratories indicate that great progress in the field of toxicology is to be expected from the development of the functional genomics technologies. Especially in safety testing, with the importance of a complete monitoring of possible adverse effects induced by a (new) compound on the complete organism, we see that functional genomics approaches will prove to be indispensable.

Acknowledgements

The authors wish to thank Dr. F. Schuren, Dr. T. van der Lende, E. Wesseling and M. Havekes for expert-assistance with the cDNA microarray technology, and M. van den Wijngaard for assistance in sample-isolation. Dr. J. van Nesselrooij supported the metabonomics measurements, and Dr .J. Groten, Dr. B. Lake (BIBRA UK) the case study on the hepatotoxicity by food additives. Dr. M. Slijper and M. Damen (Dept. of Biomolecular Mass Spectrometry, Utrecht University) helped with the protein mass spectrometry analysis. We would like to thank Dr. A. Conesa and M. van Erk for helpful discussions on data analysis.

Internet references

Swissprot	http://us.expasy.org/
European Bioinformatics Institute (EBI)	http://www.ebi.ac.uk/
Expression Profiler at the EBI	http://ep.ebi.ac.uk/EP/EPCLUST
NCBI's UniGene,Genbank, OMIM, etc All from	http://www.ncbi.nlm.nih.gov/
Swissprot	http://us.expasy.org/
GeneCards	http://bioinfo.weizmann.ac.il/cards/
KEGG pathway maps	http://www.genome.ad.jp/kegg/
Biocarta pathway maps	http://www.biocarta.com/genes/index.asp

References

[1] J. L. DeRisi *et.al.* Exploring the metabolic and genetic control of gene expression on a genomic scale, Science, 27824-10-1997, pp. 680-686

[2] M. Schena *et.al.* Quantitative monitoring of gene expression patterns with a complementary DNA microarray, Science, 27020-10-1995, pp. 467-470

[3] A. Shevchenko *et.al.* Linking genome and proteome by mass spectrometry: large-scale identification of yeast proteins from two dimensional gels, Proc.Natl.Acad.Sci.U.S.A, 9310-12-1996, pp. 14440-14445

[4] K. P. Gartland *et.al.* Application of pattern recognition methods to the analysis and classification of toxicological data derived from proton nuclear magnetic resonance spectroscopy of urine, Mol.Pharmacol., 391991, pp. 629-642

[5] J. K. Nicholson *et.al.* 'Metabonomics': understanding the metabolic responses of living systems to pathophysiological stimuli via multivariate statistical analysis of biological NMR spectroscopic data, Xenobiotica, 291999, pp. 1181-1189

[6] J. D. Robertson and S. Orrenius Molecular mechanisms of apoptosis induced by cytotoxic chemicals, Crit Rev.Toxicol., 302000, pp. 609-627

[7] E. Holmes *et.al.* Metabonomic characterization of genetic variations in toxicological and metabolic responses using probabilistic neural networks, Chem.Res.Toxicol., 142001, pp. 182-191

[8] A. Brazma and J. Vilo Gene expression data analysis, FEBS Lett., 48025-8-2000, pp. 17-24

[9] M. R. Fielden *et.al.* In silico approaches to mechanistic and predictive toxicology: an introduction to bioinformatics for toxicologists, Crit Rev.Toxicol., 322002, pp. 67-112

[10] E. F. Nuwaysir *et.al.* Microarrays and toxicology: the advent of toxicogenomics, Mol.Carcinog., 241999, pp. 153-159

[11] S. Farr and R. T. Dunn Concise review: gene expression applied to toxicology, Toxicol.Sci., 501999, pp. 1-9

[12] S. J. Bulera *et.al.* RNA expression in the early characterization of hepatotoxicants in Wistar rats by high-density DNA microarrays, Hepatology, 332001, pp. 1239-1258

[13] J. F. Waring *et.al.* Clustering of hepatotoxins based on mechanism of toxicity using gene expression profiles, Toxicol.Appl.Pharmacol., 17515-8-2001, pp. 28-42

[14] J. F. Waring *et.al.* Microarray analysis of hepatotoxins in vitro reveals a correlation between gene expression profiles and mechanisms of toxicity, Toxicol.Lett., 12031-3-2001, pp. 359-368

[15] H. K. Hamadeh *et.al.* Gene expression analysis reveals chemical-specific profiles, Toxicol.Sci., 672002, pp. 219-231

[16] M. E. Burczynski *et.al.* Toxicogenomics-based discrimination of toxic mechanism in HepG2 human hepatoma cells, Toxicol.Sci., 582000, pp. 399-415

[17] J. P. Groten *et.al.* An analysis of the possibility for health implications of joint actions and interactions between food additives, Regul.Toxicol.Pharmacol., 312000, pp. 77-91

[18] W. H. Heijne *et.al.* Toxicogenomics of bromobenzene hepatotoxicity: a combined transcriptomics and proteomics approach, Biochem.Pharmacol., 651-3-2003, pp. 857-875

[19] H. Thor *et.al.* Biotransformation of bromobenzene to reactive metabolites by isolated hepatocytes, Adv.Exp.Med.Biol., 136 Pt A1981, pp. 287-299
[20] A. F. Casini *et.al.* Liver glutathione depletion induced by bromobenzene, iodobenzene, and diethylmaleate poisoning and its relation to lipid peroxidation and necrosis, Am.J.Pathol., 1181985, pp. 225-237
[21] T. J. Monks *et.al.* Bromobenzene and p-bromophenol toxicity and covalent binding in vivo, Life Sci., 308-3-1982, pp. 841-848
[22] S. S. Lau and T. J. Monks The contribution of bromobenzene to our current understanding of chemically-induced toxicities, Life Sci., 421988, pp. 1259-1269
[23] N. E. Miller *et.al.* Bromobenzene metabolism in vivo and in vitro. The mechanism of 4-bromocatechol formation, Drug Metab Dispos., 181990, pp. 304-308
[24] Y. M. Koen *et.al.* Identification of three protein targets for reactive metabolites of bromobenzene in rat liver cytosol, Chem.Res.Toxicol., 132000, pp. 1326-1335
[25] F. Oesh and M. Arand Xenobiotic Metabolism,1st1999, pp. 83-110
[26] R. S. Friling *et.al.* Xenobiotic-inducible expression of murine glutathione S-transferase Ya subunit gene is controlled by an electrophile-responsive element, Proc.Natl.Acad.Sci.U.S.A, 871990, pp. 6258-6262
[27] Y. Tsuji *et.al.* Coordinate transcriptional and translational regulation of ferritin in response to oxidative stress, Mol.Cell Biol., 202000, pp. 5818-5827
[28] P. S. Guzelian and N. A. Elshourbagy Induction of hepatic heme oxygenase activity by bromobenzene, Arch.Biochem.Biophys., 1961979, pp. 178-185
[29] N. Hill-Kapturczak *et.al.* Heme oxygenase and the kidney, DNA Cell Biol., 212002, pp. 307-321
[30] A. F. Suffredini *et.al.* New insights into the biology of the acute phase response, J.Clin.Immunol., 191999, pp. 203-214
[31] S. U. Ruepp *et.al.* Genomics and proteomics analysis of acetaminophen toxicity in mouse liver, Toxicol.Sci., 652002, pp. 135-150

Toxicogenomics and Proteomics
J.J. Valdes and J.W. Sekowski (Eds.)
IOS Press, 2004

The Toxicogenomics of Low-Level Exposure to Organophosphate Nerve Agents

Jennifer Weeks SEKOWSKI[1], Maryanne VAHEY[2], Martin NAU[2], Mary Anne OREHEK[1], Stephanie MAHMOUDI[1], Jennifer BUCHER[1], Jay HANAS[3], Kevin O'CONNELL[1], Akbar KHAN[1], Mike HORSMON[1], Darrel MENKING[1], Christopher WHALLEY[1], Bernard BENTON[1], Robert MIODUSZEWSKI[1], Sandra THOMSON[1], and James J. VALDES[1]

[1]*Edgewood Chemical Biological Center, U.S. Army RD& E Command, Aberdeen Proving Ground, MD, US*
[2]*Walter Reed Army Institute of Research, Rockville, MD, US*
[3]*University of Oklahoma Health Sciences Center, Oklahoma City, OK, US*

Abstract. To date, there is a great paucity of information regarding the molecular genetic effects of low-level, but militarily significant, exposures to chemical warfare agents. Gene expression, the process whereby genes are transcribed from the DNA template into messenger RNA (mRNA) and then translated into the gene product (e.g., polypeptide or protein), is a highly regulated and sensitive cellular event. The internal signals that trigger induction (or repression) of gene transcription can respond within seconds to a toxic insult. Furthermore, this process is sensitive to the degree that alteration of gene expression can occur at toxicant exposure levels far below that which cause physiological symptoms. Most importantly, changes in gene expression have been demonstrated to precede toxicant-associated injury or disease by weeks, months, and even years. The advent of high quality DNA microarray technology has made it possible to measure the expression level of thousands of genes simultaneously at a given time and/or dose point in a cell or tissue-based system. The use of new functional genomic techniques (such as DNA microarrays) to study the response of genes to toxic insults has created a new sub-field of toxicology called toxicogenomics. The use of toxicogenomic approaches to study low-level exposure to chemical nerve agents and other militarily significant chemicals is anticipated to greatly advance the understanding of the molecular genetic underpinnings of these types of toxic exposures and lead to the development of better detection, protection and decontamination measures.

1. Introduction

1.1 Background on Organophosphate Nerve Agent Toxicology

Over the past more than 50 years, much has been learned about the acute effects of organophosphate (OP) nerve agents. The most common OP agents are the "G" series agents and VX. The G agents were synthesized and developed by the Germans just prior to World War II. Initially synthesized as pesticides, they were found to be acute inhibitors of the enzyme Acetylcholinesterase (AchE) in humans, causing death within minutes from accumulation of synaptic Acetylcholine. Soon after the war ended, the U.S. and its allies discovered the stockpiles, and began synthesizing and testing the nerve agents. While the only known battlefield use of nerve agents was in the Iran –Iraq War, many countries as well as non-state sponsored political groups have possession of and knowledge to manufacture organophosphate nerve agent and nerve agent munitions.

Since the Gulf War, there has been increasing interest in understanding the possible longer lasting effects of low-level nerve agent exposure. In the U.S., this new research focus, led primarily by the U.S. Veterans Administration, has engendered much of the recent work in the area of low-level agent toxicology. The terrorist attacks in Japan in the mid-1990's raised much awareness about the lack of information regarding the possibility of mass exposure to low-level chemical agent plumes. The Aum Shinrikyo cult of Japan used sarin in the two terrorist attacks. The first attack was in an apartment complex in Matsumoto, Japan (in 1994) and the second (1995) was in the Tokyo subway. In these attacks, 5,500 people were treated and 13 people died. Many of those treated were exposed to low-levels of sarin vapor (Sidell et al., 2002). From these tragic events, however, at least one positive outcome has been (and continues to be) gleaned. The experience of the patients and the physicians who treated them are able to give science and medicine new, invaluable insights into the human response to low-level sarin exposure. Hopefully, the operational insight and the overall the greater awareness of the problem will lead to the development of better detection, protection, and treatment measures in the near future.

Currently, we are attempting to understand, using cell culture and model animal systems, what molecular events occur during and following low-level exposure to organophosphate nerve agents, particularly GB (sarin) and GF (cyclosarin). By correlating the molecular level alterations with changes in organ-level biochemistry, we hope to answer questions that will spur the development of the best detection, decontamination, and protective measures possible. In doing so, we hope to prevent the future suffering of both military and civilian lives.

1.2 Properties and Mechanisms of Action of Sarin and Cyclosarin

Sarin (GB; isopropyl methylphosphonofluoridate, $CH_3PO(F)OCH(CH_3)_2$) is an acutely toxic, odorless (when in the pure state) organophosphate nerve agent. It has a relatively high volatility, (approximately that of water) making it relatively non-persistent in materials that have absorbed the liquid and vapors. It also undergoes relatively rapid hydrolysis out side of the pH range 4-7, resulting in the production of two less toxic compounds, hydrogen fluoride and methyl phosphonic acid. Cyclosarin (GF; (cyclohexyl methylphosphonofluoridate; $CH_3PO(F)OC_6H_{12}$) is just as acutely toxic as sarin, but is approximately 38 times less volatile than sarin at 25°C (GF = 581 mg/m^3; GB= 22,000 mg/m^3). It also is reported to have a slightly sweet odor resembling peach, musk or shellac.

The primary toxic mechanism of action for both compounds is the inhibition of butyrylcholinesterase in serum and acetylcholinesterase in red blood cells and at cholinergic receptors in tissue. After either cholinesterase enzyme is bound by the agent, it is unable to hydrolyze the neurotransmitter acetylcholine, resulting in accumulation of acetylcholine at the receptors. This causes hyperstimulation of both muscarinic and nicotinic acetylcholine receptors. Symptoms of hyperstimulation at muscarinic receptors include miosis (pinpointed pupils), eye pain, respiratory distress (tightness of chest and dyspnea), exocrine gland stimulation (tearing, sweating, salivation), GI disturbances (nausea, vomiting, cramping, diarrhea), and bradycardia. Stimulation of the nicotinic receptors causes a range of effects including twitching, convulsions, peripheral and respiratory muscle weakness, flaccid or rigid muscle tone, and, ultimately, paralysis. Sarin and cyclosarin have also been reported to demonstrate non-cholinergic effects, such as disturbing other enzymatic activities, altering other neurotransmitter systems, and binding directly to other receptors. It is known that certain aspects of the nervous system (CNS) are directly affected by sarin and cyclosarin exposure. CNS effects include headache, drowsiness, giddiness, excessive dreaming, fatigue, forgetfulness, anxiety, dyspnea, cyanosis, and death. For all of the above, the type and severity of the symptoms depend on the level and duration of exposure and the route(s) of exposure. Despite more than 50 years' worth of research, large data gaps exist in the body of OP agent literature. Particularly lacking is consensus regarding the near- and long-term clinical significance of the non-cholinergic effects and the overall effects of acute and chronic low-level exposure.

1.3 Known Effects of Chronic Low-Level s and Cyclosarin Exposure

The largest hindrance in reaching consensus in understanding the effects of low-level exposure is the lack of human data from well-designed, controlled studies. Researchers have been able to draw certain conclusions, however, from several studies that used very short-term exposures of OP agents on humans, from several accidental high-dose human exposures, and from studies using animal models. There is also a considerable body of literature regarding the effects of OP insecticides. While many of the data are relevant, it is dangerous to draw too close a parallel to OP nerve agents. The behavior and toxicology of the various OP compounds vary widely between the particular compounds. Overall, the studies that have been carried out to examine the effects of chronic exposure to OP compounds, have revealed many general insights toward understanding the effects of OP nerve agents on the various physiological systems of the body.

1.4 Organophosphate-Induced Delayed Polyneuropathy (OPIDN)

The most known notable area of research in assessing whether chronic low-level sarin exposure can cause lasting neuronal damage is the syndrome known as organophosphate-induced delayed neuropathy (OPIDN). For more than 100 years, acute OP compound exposure has been linked to the development of OPIDN. The neuropathy generally presents as muscle weakness and ataxia, especially in the lower limbs. This weakness typically occurs between 8 and 14 days after an acute exposure, depending on the type of compound and route and extent of exposure. Occasionally, the neuropathy progresses in severity toward true paralysis. The neuropathy is associated with demyelination of peripheral nerve axons and certain parts of the central nervous system (CNS) (Lotti, 1992). It is commonly believed that the demyelination relies on the OP compound's ability to phosphorylate and irreversibly inhibit the enzyme neuropathy target esterase (NTE)(Johnson, 1982). The role NTE plays in the initiation of OPIDN remains unclear. The majority of the OPIDN studies have been carried out using the hen, rat or mouse as a model. In two studies using humans exposed to symptomatic doses of GB (Munro, 1994), no findings of OPIDN were reported. Thus, given the human data available, it does not appear that humans develop OPIDN from GB exposure, even at acute doses.

1.5 Other Neurological Changes

Many other symptoms of chronic OP exposure are even less well defined than OPIDN. These include behavioral and psychological changes, deficits in memory and cognition.

Reports of changes in behavior and psychological state after exposure to organophosphates have been around for more than fifty years (Buccofusco, 1998). Some of the manifestations reported included altered EEG measures, reduced concentration and memory, reduced information processing and psychomotor speed, linguistic deficits, depression, anxiety, sleep disturbance, and easy fatigability (Burchfiel and Duffy, 1983; Ecobichon and Joy, 1994; Sidell et al., 2002). The existing body of data deals primarily exclusively with levels of exposure that cause symptoms. As such, it is likely that most these reported changes are due primarily to inhibition of ChE, and the neurochemical imbalances that result from it. However, due to the extreme amount of individual variability in the type, degree, and duration of the physiological response to an OP exposure, it is not outside of the realm of possibility that these symptoms could occur in individuals exhibiting few or no other signs of OP exposure (Sidell et al., 2002; Ecobichon and Joy, 1994). Overall, there is very little direct and convincing data suggesting that low levels of GB exposure lead to consistent EEG, psychological, or behavioral alterations. More research is needed in this area.

1.6 General Health Effects

Other pathological effects of low-level OP exposure have been investigated over the past 50 years, yet very little additional data has been revealed. Carcinogenicity and genotoxicity have been measured in many studies using high doses of OP compounds, resulting in no findings of either pathogenic mechanism. OP exposures have also been assessed for teratogenicity and other reproductive effects, yet potential involvement of either has not been demonstrated.

2. TOXICOGENOMICS - Moving Beyond Classical Organophosphate Toxicology

The body of organophosphate toxicological research revealed over the past fifty or more years is predominantly classical toxicology. The primary goal of these toxicological investigations has been and continues to be to measure physical parameters (e.g., overt symptoms of exposure, physical activity, enzymatic activity), histological features of tissue samples (histopathology), and blood chemistry (e.g., ChE measures) as indicators to identify adverse effects of exposure. These data have been and continue to be very useful diagnostic indicators of acute levels of OP exposure, but they are less useful for understanding the molecular mechanisms of toxicity or to gain insight into the potential for longer-term impacts of OP exposure. In addition, classical toxicological methods are often not sufficiently sensitive to detect very low-level toxicity. Tools and methods that evaluate toxicity and pathology at the mechanistic level are essential in order to establish a meaningful estimates for molecular-level lowest observable adverse effect levels (LOAELs) for organophosphate nerve agents. Of the tools available, one of the most exciting and useful ones is the DNA microarray. A microarray allows one to monitor the expression of tens to thousands of genes simultaneously. Microarray- based investigations have already begun to make an impact on the field toxicology. A toxicogenomic approach to a question that includes DNA microarray-generated gene expression data that is verified using RT-PCR or another technique, and is complemented with an investigation into protein expression (proteomics) begins to build a compelling foundation for the discovery of new molecular toxicological mechanisms.

2.1 Gene Expression Profiling

DNA microarray-initiated research has been used in the past several years to discover new molecular mechanisms of toxicity, but also to establish gene expression profiles (or signatures) that can reliably reflect the molecular status or phenotype of cell and tissues. By building databases of these signatures, it is possible to build test sets that can be used to identify, by their gene expression profile, samples of unknown composition or disease state, or that have been exposed to unknown compounds. A terrific example of this possibility is demonstrated by the laboratory of Golub *et al.* (1999) . By clustering gene expression profiles from acute myeloid leukemia (AML) and acute lymphoblastic leukemia (ALL) cells, they were able to correctly classify 29 out of 34 bone marrow or peripheral blood samples as being either AML or ALL. Similarly, a study carried out by Ross *et al.* (2000) demonstrated the ability to accurately classify 60 diverse types of tumor cell lines based solely on gene expression profile.

2.2 Use of Gene Expression Profiling in Toxicology

Several studies have also been published recently that show that gene expression profiles from cells treated with a toxicant or drug treatment are indicative of the mechanism of action of the xenobiotic. A study conducted by Scherf et al. (2000) demonstrated that gene expression profiling could indeed correctly classify anti-cancer drugs' mechanism of action. The cellular responses (measured by gene expression profiles) to the 118 anti-cancer therapeutics clustered accurately according to the mechanism of action of the therapeutic agent to which the cell type was exposed, regardless of the cell type. Another study, by Waring et al. (2001), addressed the question of whether gene expression profiling could be used to predict a toxic response. Using rats exposed to a battery of 15 known hepatotoxins, they successfully demonstrated a strong correlation between the rat liver gene expression profiles, histopathogical profiles, and clinical chemistry. Overall, these two studies make a compelling case for the utility of gene expression profiling to predict the mechanism of toxicity of exogenous compounds.

3. DEVELOPMENT OF THE OP NERVE AGENT DNA MICROARRAY

3.1 Objective- The objective of this work is to develop a DNA microarray containing the genetic expression signatures of chemical warfare agent(s). The resulting "CWA Signature DNA Microarray" will enhance CWA detection and protect the soldier in <u>three important ways</u>: 1) it will provide a "before and after" genetic snapshot of soldiers who enter a potentially contaminated environment and give unambiguous verification of exposed individuals; 2) identification of genes altered by CWA exposure may aid in the improvement of preventive or therapeutic measures, and 3) it will provide a means to help identify individuals illegally manufacturing or working with CWA. The "CWA Signature DNA Microarray" will be used in the laboratory to screen CWAs, mixtures of CWAs and TICs, and other potential toxicants for their mechanism of action and similarity to other CWAs and toxicants. It is anticipated that data obtained through the development and use of this assay will fill many critical data gaps in support of developing CB defensive detection, protection, and decontamination materiel. Overall, this long-range project is expected to contribute tremendously to the missions of the Low-Level Operational Toxicology and the Deployment Toxicology Programs.

3.2 Approach- We obtain low-level CWA (i.e. GB and GF) exposed rat blood, brain and liver tissue through an ongoing collaboration with the Toxicology Team at ECBC (Drs. Mioduszewski, Whalley, and Benton). The toxicology team at ECBC is world renowned for their whole body (rat) inhalation model of chemical agent exposure. This exposure model utilizes the most widely accepted mammalian model in the field of toxicology, provides the most relevant route of low-level chemical agent exposure and provides extremely accurate dosing information.

With the advent of high-quality, commercially available gene expression arrays, array technology has become a widely accessible functional genomics tool for the molecular toxicologist. The transcript processing and hybridization is carried out using themethod as described by the Vahey laboratory (Vahey *et al.*, 2002). Briefly, mRNA is extracted from the tissue, is converted to cDNA, and then to fluorescently-labeled cRNA. The labeled cRNA is hybridized to the DNA array, stained, and then visualized using a fluorometer. The raw expression level data is translated using bioinformatics software and databases into meaningful biological information such as relative induction / repression ratios, complex pattern analyses, association with biochemical pathways, and participation in disease processes. We have chosen to use the Rat Toxicology U34 and the Rat Whole Genome U34A Gene Chip from Affymetrix in our investigations of GB and GF in the rat model. The Rat Neurobiology Gene Chip, RN-U34, contains more than 1200 known genes and expressed sequence tags (ESTs), and the Rat Genome Gene Chip, RG-U34A, contains more 6,000 known genes and ESTs. The genes included in the arrays encode polypeptides critical for cellular functions such as stress response, apoptosis, DNA replication/ repair, recombination, transcription, cell-cell communication, and for such structures as ion channels and cell surface receptors, to name a few. Of the thousands of genes on either array, only a small subset are significantly (greater than 2 fold) induced or repressed compared to the control animals. Once verified by RT-PCR, the significantly altered genes will be included in our database and considered for integrated in the design of the CWA Signature DNA Microarray.

3.3 Preliminary Data- To date, we have analyzed and compiled significant data from our gene expression studies on chlorpyrifos and GB. Our initial pilot study revealed that rats injected with an LD30 dose of the organophosphorus (OP) pesticide, chlorpyrifos, had significantly altered gene expression patterns compared to the control animals (Fig 1). Furthermore, we observed that several of the genes remained altered at 24 hours post exposure. The arrows in Fig.1 indicate that the expression level for the glutamate aspartate transporter (GLAST) increased to more than four fold above the level observed in the control condition. Overall, the pilot study demonstrated the utility of the microarray approach and demonstrated to our satisfaction that gene expression analysis could provide a valuable window through which we could uncover genetic mechanisms of OP compound toxicity (Fig. 2) (Sekowski *et al.*, 2001; Sekowski *et al.*, 2003). Our more recent studies, examining the toxicogenomic effects of GB (sarin) and GF (cyclosarin) whole body inhalation exposure in rats, have resulted in a more significant body of data. Not only do the more recent studies contain statistically meaningful numbers of animals, but also the animals have been exposed to actual OP nerve agents via an operationally relevant route of exposure. Thus, we were able to collect a tremendous amount of statistically significant gene expression data, and begin to reveal some potential toxic mechanisms of action that appear to occur even at these low levels of CWA exposure (Sekowski, *et al.*, 2003B).

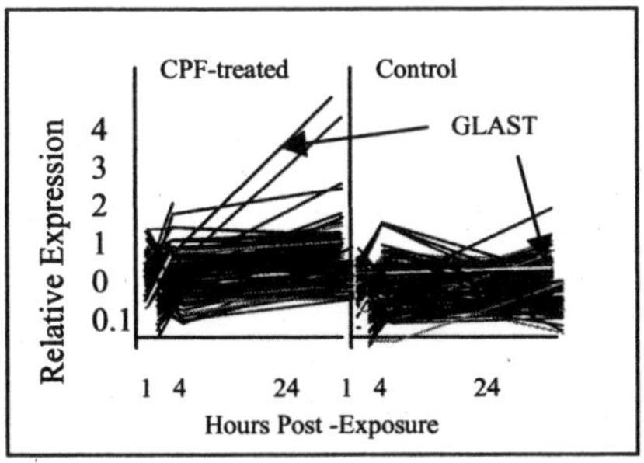

Figure 1

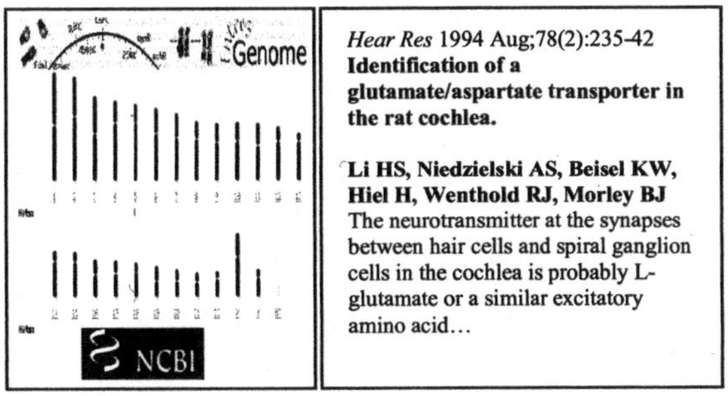

Figure 2

Significance of our gene expression data is based upon several layers of metrics. First, the overall quality of the data is vetted. Using Affymetrix's software tool, Microarray Suite, the raw image of the processed GeneChip is examined for flaws, bubbles, etc. Next, the background noise, consistency in scaling factors, and the sets of positive (BioB, BioC, GAPDH, etc) control genes are examined. If the data looks consistent within the experiment and is of high quality, the data is transferred to other bioinformatics platforms. Currently, we are using Gene Spring from Silicon Genetics and Data Ming Tool from Affymetrix. Using these programs, the data is run through statistical tests such as Mann-Whitney (Wilcoxin Sign Rank Test) and ANOVA in order to demonstrate the consistency within and difference between sets of data. The data is also easily visualized using either one of these programs. Trends and outliers are easily picked out. Clustering analysis (e.g., K-means and functional) and PCA can be performed on the data. These software tools have allowed gene lists to be honed from thousands of genes to just a handful of genes that are significantly altered.

By measuring gene expression in the brain, liver and blood from both male and female animals exposed to different doses and durations of GB and GF vapor we have observed several important trends. The first is that significant differences in gene expression can be measured even after a one-week "recovery period" following exposure. This is a highly significant finding, and has exciting implications for the persistence of gene expression signatures in the body. The second important trend observed was that there appears to be a dose–response relationship to the level of OP agent exposure and the range and average of gene expression changes (Fig. 3-4).

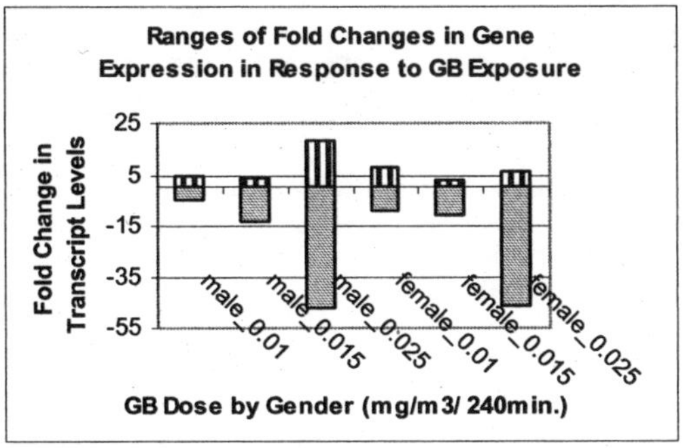

Figure 3. Dose-response Relationship of GB Exposure and Range of Gene Expression Changes. The ranges of fold changes in gene expression increased with dose, suggesting a general dose- response in gene expression.

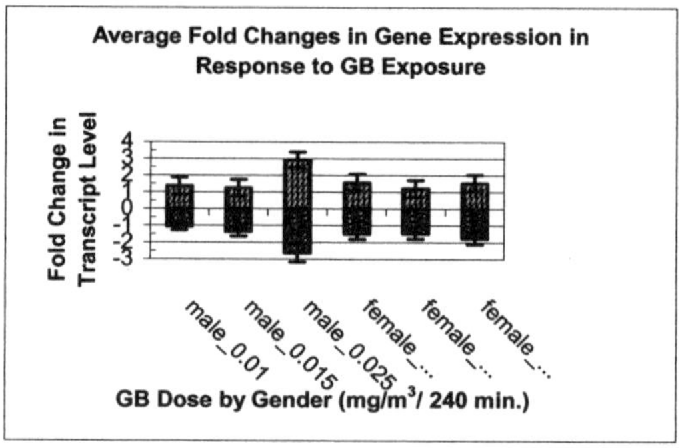

Figure 4. Dose-response Relationship of GB Exposure and Average Fold Change in Gene Expression. The average fold changes also increased slightly with dose. This trend is more pronounced in the male rats.

The third trend observed to date has been that there appears to be significant differences in the gene expression patterns in the males versus the female rats following GB exposures (Fig 5). This finding, together with the finding from the Toxicology Team that the females are nearly twice as sensitive to the miotic effects of GB (Mioduszewski *et al.*, 2002), suggests that the response to certain CWAs may be different in men and women. Thus, all future studies will incorporate female animals as well as males.

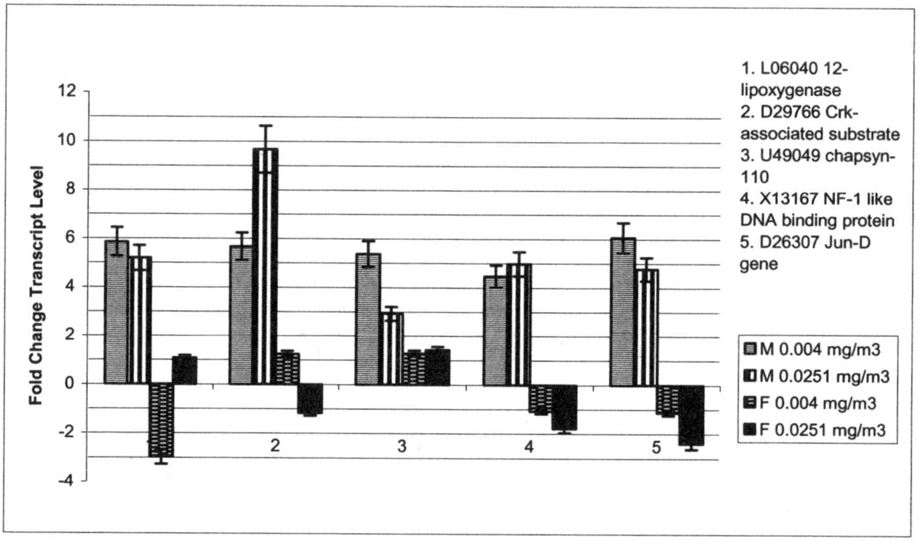

Figure 5. Male-Female Differences in Brain Gene Expression After Low-Level GF Exposure. Gene expression of female rats exposed to low-level GB vapor for 4 hours showed markedly less response or a weak opposite response in gene expression compared to males. Further investigation into the functions and interrelatedness of these particular genes is ongoing.

The most recent data in this study surround the identity of the genes whose expression was significantly altered after exposure GB and GF. The gene expression investigation using the animals exposed to GB utilized the Rat Neurobiology Gene Chip. This GeneChip is comprised of a 1,324 gene and EST subset of the whole rat genome. These are genes that Affymetrix has designated as related to neuronal processes. We were able to glean many promising leads from the GB-exposed rat brain RNA using this array. Figure 6 represents a sampling of the highest upregulated genes that were detected using the array. Lesser up-

and down-regulated genes may also be important functionally (and may be investigated later), but for the sake of brevity only the data displayed in Figure 6 will be discussed in this proposal. In Figure 6, for example, it is apparent that the largest change in gene expression was measured for the males at the 0.0.025 mg/m3 level. The authors of this proposal doubt that this greatly enhanced response occurring near the ECT_{50} (for miosis for a 4 hour GB vapor whole body inhalation exposure in male Sprague-Dawley rats) is coincidental. The ECT_{50} (miosis) was reported to be approximately 0.012 mg/m3 for females and 0.024 mg/m3 for males (Mioduszewski *et al.*, 2002). Thus, the large change in gene expression at 0.024 mg/m3 in males is likely reflecting some of the neurochemical changes that comprise the environment in which miosis occurs. However, it is not clear why the females do not show this same trend in gene expression at their ECT_{50} (near 0.012 mg/m3).

Although this graph (Fig. 6) is rather large, it shows only the genes most significantly altered in expression after the rats' exposure to GB. In this abbreviated list there are several ion and amino acid receptor proteins including: two potassium channel proteins (X62839, U69882), a sodium channel protein (M22253), the P2X2-5 receptor (ion-gated purinergic receptor)(AF020758), a glutamate receptor (M38061), and a putative ion channel protein (AF078779). Also upregulated, synapsin III (AF056704) is a synaptic vesicle protein essential for normal Ca2+ regulation of neurotransmitter release. An upregulation of the channels proteins and the synaptic vesicle protein at one week after exposure could signify that the brain is still restoring homeostasis after the injury caused by GB-exposure. The induction of the neurotransmitter induced early gene 2 (part of the ania family of genes)(AF03008) is know to be stimulated within minutes of neurotransmitter receptor activation. This class of genes encodes transcription factors that regulate late response genes. This class of gene has been implicated in response to hypoxic and other types of brain injury (Raghavendra *et al.*, 2003). Up regulation may signify continuing recovery from neuronal injury. Similarly, a few cell damage repair and plasticity-associated proteins were also found to be upregulated. These include the neurodegeneration-associated protein (involved in repair/plasticity)(D32249), neuroaxin (involved with neuron-microtubule associations)(C16623), neuron specific protein (PEP19)(highest in cerebellum)(M24852), and neuronal protein 25 (NP25)(highest in cerebellum; related to calponin) (M84725). We found also two inflammation/ immune-related genes: CD44 (a lymphocyte stimulating protein)(AFO14365) and vascular cell adhesion molecule-1 (a member of the immunoglobulin super family and an important factor involved in the process of leukocyte recruitment in inflammation) (M84488).

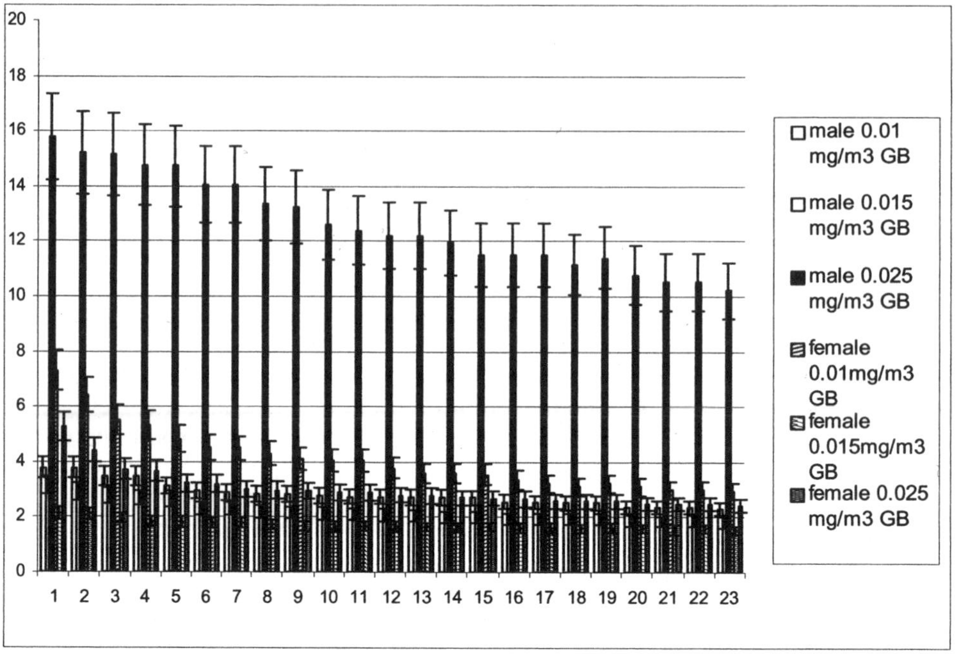

Figure 6. Gene Expression Upregulation One Week After Low-Level GB Exposure. A partial list of highly significant ($p<=0.001$) gene expression alterations remaining after a one week recovery period following exposure to GB vapor at concentrations near the levels which caused miosis in these animals.

Key to gene in Fig. 6:

1	rc_AI227660 EST224355
2	M84488 Rat vascular cell adhesion molecule-1
3	rc_AI230404 EST227099 R
4	AF030087UTR#1 neurotransmitter-induced early gene 2 (ania-2)
5	U88324 RNU88324 G protein beta1 subunit (rGb1)
6	AF020758 P2X2-5 receptor (P2X2)
7	AF044910 survival motor neuron (smn)
8	U31554 limbic system-associated membrane protein
9	rc_AI101255 EST210544
10	X62839 potassium channel protein (3120 bp)
11	D84450 Na+,K+-ATPase beta-3 subunit
12	M24852 neuron-specific protein PEP-19
13	U69882 RRU69882 calcium-activated potassium channel rSK2 (SK)
14	AF078779 putative four repeat ion channel
15	J05510 inositol-1,4,5-triphosphate receptor
16	M84725 neuronal protein (NP25)
17	X16623 neuraxin
18	U03470 RFas antigen ligand
19	rc_AI237836 EST234398
20	AF014365 CD44
21	AF056704 synapsin IIIa
22	D32249 RATNDAP1 neurodegeneration associated protein 1
23	M38061 glutamate receptor (GluR-B)

Lastly, the down selection of the significant genes that appear to be the most interesting or most promising of revealing biochemical mechanism appear are currently being verified using real time reverse transcriptase (RT-PCR). RT-PCR is the most widely used method to verify gene expression array results. To date, we have optimized approximately 30 primer sets for the PCR reaction conditions. This work has been re-optimized using the more accurate and sensitive PCR method using TaqMan primers and probes. At the writing of this manuscript, we have some preliminary data that suggests that this new approach is working (Fig. 7).

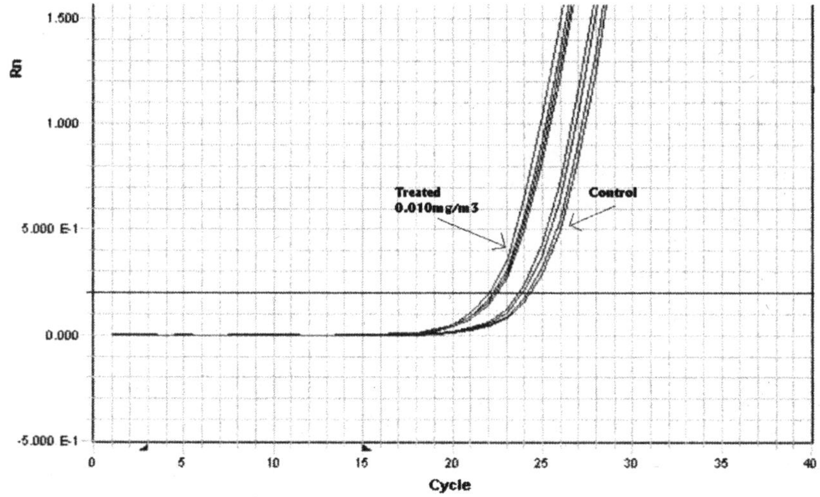

Figure 7. RT-PCR of the transcript for (U53927) brain astroglial high-affinity cationic amino acid transporter (RCAT2) from male rat brain exposed to GB. The CT_{50} for the exposed male rats (0.01 mg/m3 for 4 hours) was 22.5 and the controls were 24.5 cycles. A difference of 2 cycles translates to a 4-fold difference in transcript level. While the Rat Neurobiology GeneChip measured the transcript level as a 7-fold upregulation, the fold level change is considered in the same range as that measured by RT-PCR.

References

Buccofusco, J.J. (1998) Chronic Organophosphorus Exposure and Cognition. Govt Reports Announcements & Index (GRA&I), Issue 06, 1998.

Burchfiel, J.L. and Duffy, F.H. (1983) Organophosphate neurotoxicity: chronic effects of sarin on the electroencephalogram of monkey and man. Neurobehav Toxicol Teratol., 4(6):767-78.

Ecobichon, D.J. and Joy, R.M. (1994) Pesticides and Neurological Diseases, 2nd Edition. CRC Press, Inc: Boca Raton, FL.

Golub T.R., Slonim, D.K., Tamayo, P., Huard, C., Gaasenbeek, M., Mesirov, J.P., Coller, H., Loh, M.L., Downing, J.R., Caligiuri, M.A., Bloomfield C.D., Lander, E.S. (1999) Molecular classification of cancer: class discovery and class prediction by gene expression monitoring. Science, 286(5439):531-7.

Johnson, M.K. (1982) The Target For Initiation Of Delayed Neurotoxicity By Organophosphorus Esters: Biochemical Studies And Toxicological Applications Reviews in Biochemical Toxicology, Vol.4:141-212.

Lotti, M. (1992) The pathogenesis of organophosphate induced delayed polyneuropathy. Crit. Rev. Toxicol., 21: 465-487.

Mioduszewski R.J., Manthei, J.H., Way, R., Burnett, D.C., Gaviola, B.P., Muse, W.T., Sommerville, D.R., Crosier, R.B., Scotto, J.A., McCaskey, D.A., Crouse, C.L., Matson, K.L. and Thomson, S.A. (2002) Low-Level Sarin Vapor Exposure in Rats: Effect of Exposure Concentration and Duration on Pupil Size. U.S. Army Technical Report, ECBC-TR-235, May 2002.

Munro, N.B., Ambrose, K.R., and Watson, A.P. (1994) Toxicity of the Organophosphate Chemical Warfare Agents GA, GB, and VX: Implications for Public Protection Environmental Health Perspectives, Vol. 102 (1): 18-38.

Raghavendra Rao, V.L., Dhodda, V.K., Song, G., Bowen, K.K., Dempsey, R.J. (2003) Traumatic brain injury-induced acute gene expression changes in rat cerebral cortex identified by GeneChip analysis. J Neurosci Res., 71(2): 208-19.

Ross, D.T., Scherf, U., Eisen, M.B., Perou,C.M., Rees, C., Spellman, P., Iyer, V., Jeffrey, S.S., Van de Rijn, M., Waltham, M., Pergamenschikov, A., Lee, J.C., Lashkari, D., Shalon, D., Myers, T.G., Weinstein, J.N., Botstein, D., Brown, P.O. (2000) Systematic variation in gene expression patterns in human cancer cell lines. Nature Genetics, 24(3):227-35.

Scherf ,U., Ross, D.T., Waltham, M., Smith, L.H., Lee, J.K., Tanabe, L., Kohn, K.W., Reinhold, W.C., Myers, T.G., Andrews, D.T., Scudiero, D.A., Eisen, M.B., Sausville, E.A., Pommier, Y., Botstein, D., Brown, P.O., Weinstein, J.N. (2000) A gene expression database for the molecular pharmacology of cancer. Nature Genetics, 24(3):236-44.

Sekowski, J.W., O'Connell, K.P., Khan, A.S., Vahey, M., Nau, M., Kahlil, M., Eldefrawi, M.E., and Valdes, J.J. (2001) Changes in gene expression after exposure to the organophosphorus insecticide chlorpyrifos. U.S. Army Technical Report, ECBC-TR-210, September 2001.

Sekowski, J.W, O'Connell, K.P., Khan, A.S., Vahey, M., Nau, M., Kahlil, M., Eldefrawi, M.E., and Valdes, J.J.. "Changes in Gene Expression After Exposure to an Organophosphorus (OP) Agent". (2003) In Proceedings from "Alternative Methods in Toxicological Methods for the New Millenium: Science and Application". NIH, Bethesda, MD., pp. 413-427.

Sekowski, J.W, Vahey, M., Nau, M., Orehek, M.A., Mahmoudi, S.F., Bucher, J., Horsmon, M., Menking, D., Whalley, C., Benton, B., Mioduszewski, R., Thomson, S., and Valdes, J.J. (2003b) Changes in gene expression following whole body exposure to GB vapor in rats. U.S. Army Technical Report, in press.

Valdes, J.J. "Changes in Gene Expression After Exposure to an Organophosphorus (OP) Agent". (2003) In Proceedings from "Alternative Methods in Toxicological Methods for the New Millenium: Science and Application". NIH, Bethesda, MD., pp. 413-427.

Sidell, F.R., Takafuji, E.T., and Franz D.R. (2002) Medical Aspects of Chemical and Biological Warfare. Govt Reports Announcements & Index (GRA&I), Issue 11, 2002.

Vahey, M., Nau, M., Jogodzinski, J., Yalley-Ogunro, M., Taubman, N., Micjeal, N., Lewis, M. (2002) Impact of viral infection on the gene expression profiles of proliferating normal human peripheral blood mononuclear cells infected with HIV-RF. *AIDS Research and Human Retroviruses,* 18:179-192.

Waring JF, Jolly RA, Ciurlionis R, Lum PY, Praestgaard JT, Morfitt DC, Buratto B, Roberts C, Schadt E, Ulrich RG (2001) Clustering of hepatotoxins based on mechanism of toxicity using gene expression profiles. Toxicol Appl Pharmacol., 175(1):28-42.

Toxicogenomics and Proteomics
J.J. Valdes and J.W. Sekowski (Eds.)
IOS Press, 2004

Molecular Biomarkers

Soheir Saad KORRAA
Laboratory of Environmental Mutagens
National Center for Radiation Research and Technology
P.O.Box 29 Nasr City, Cairo, Egypt

Abstract. An emerging concept is that the combined action of environmental factors and individual susceptibility determines an individual's likelihood of developing cancer as well as many other aging associated diseases. Environmental factors include infectious pathogens, chemical and physical mutagens, which are frequently capable of eliciting multiple cellular changes at the genetic and epigenetic levels and can be divided into two groups: (i) DNA-damaging agents that are capable of inducing somatic mutations; and/or (ii) those that alter expression (increase or decrease) of proteins involved in multiple signal transduction pathways and induce the expression inducible proteins whose genes are highly influenced by external stimuli. Most of these elicited gene expressions are also expressed in tissue transformation and progression of tumors and are similar to the hallmarks used for cancer prognosis. Some of these proteins represent protective mechanisms against different environmental stresses, while others amplify adaptation to environmental conditions. The resultant balance between protective proteins and adaptive proteins seem to determine an individual susceptibility to develop cancer.

Introduction

Both epidemiological and experimental studies support the hypothesis that various environmental factors are implicated in the development of a wide range of malignancies [1] and other age related autoimmune diseases such as diabetes [2] and Alzheimer disease [3]. These environmental factors include chemical and physical carcinogens (e.g. cigarette smoke [4], aflatoxin B_1 [5], asbestos [6] and UV irradiation [7], ionizing radiations [8]), infectious agents (e.g. *Helicobacter pylori* [9], rhinovirus [10], schistosomiasis [11]), life-style (e.g. extensive exercise [12] and physiologic stress [13]) and dietary non-genotoxic constituents (e.g. increased fat consumption [14] and proteins [15]. Such environmental mutagens have been documented to elicit changes at both the genomic level (inducing gene mutations) and the protein level by significantly altering both basal and inducible expression of several inducible genes with or without affecting the expression of constitutive genes [16 –19]. Detectable gene mutations and alterations in signal transduction pathways together with modified post-translational proteins offer valuable molecular biomarkers for occupational and environmental human biomonitoring applied for the identification of potentially hazardous exposures before adverse health effects appear, and allow the establishment of exposure limits in order to minimize the likelihood of significant health risks [20].

Concurrently, most of these elicited gene expressions are also expressed in tissue transformation and progression of tumours and are similar to the hallmarks used for cancer prognosis. Some of these proteins represent protective mechanisms against different environmental stresses, while others amplify adaptation to environmental conditions. The resultant balance between protective proteins and adaptive proteins seem to determine an individual likelihood to develop cancer.

Definition and Types of Biomarkers

Environmental molecular biomarkers play an important role in the early detection of exposed individuals or groups at an early stage before the occurrence of significant or irreversible adverse health effects. The term "biomarker" is a general term for specific measurements of the interaction between an environmental agent and a biological system [20, 21]. In environmental epidemiology a biomarker represents a sub-clinical and reversible change, which may lead to identification of potentially hazardous exposures before adverse health effects appear, and assists in establishing the exposure limits needed for minimizing the likelihood of significant health risks. It does not represent the diagnosis but rather reflects a risk that such pathology could develop at later stage [22, 23]. Biomarkers are classified into three different types, namely biomarkers of exposure, effect and susceptibility [20]. There is an overlap between biomarkers of exposure and biomarkers of effect, because the same biomarker can be used for both measurements. Also, some of the same biomarkers are also used to measure inter-individual differences in response and thus further serve as biomarkers of susceptibility [24].

1. Biomarkers of Exposure

1.1. Definition

Biomarkers of exposure refer to measurement of the specific chemical of interest or its specific metabolite/s in a body compartment or fluid in order to identify potentially hazardous exposures before adverse health effects appear [20]. Exposure biomarkers can lead to early detection of exposure at a point where significant health effects have not occurred and consequently focus attention on external means of reducing exposure, i.e. primary prevention. They may also facilitate comparison of exposure levels in different compartments (external, blood, cellular) to identify susceptibility differences among different tissues and different individuals [24].

Identification of biomarkers of exposure, in most of the cases, depends on the concentration difference between the exposed individual and the general population [25]. It should be predictive and possess metabolic, analytical and source specificity. A biomarker is to be considered most valid if it has the ability to *predict* disease. *Metabolic specificity* refers to the extent that the analyte measured is derived from the parent chemical of interest. *Analytical specificity* refers to the capacity of the analytical method to measure exclusively the chemical that it purports to measure. *Source specificity* is of important consideration as it is very important to identify the source in order to prevent exposure [25].

A number of techniques have been developed for monitoring human exposure to environmental or occupational carcinogens. Among these methods is the measurement of the chemical itself or its metabolites in body fluids (e.g., serum or urine) or bound to cellular macromolecules such as DNA and proteins.

1.2. Macromolecular Adducts

Assessing human exposure to chemicals by the detection and quantification of adducts formed between environmental carcinogens in macromolecules such as proteins and DNA offers a reliable means of monitoring carcinogen exposure. Adducts with hemoglobin and lymphocyte DNA seem to correlate with each other and with the concentration of the mutagen.

Protein adducts such as hemoglobin adducts are advantageous in that the kinetics are well established, large amounts of globins are present in blood and no repair enzymes would remove the formed adduct once it has been generated, thus enabling possible detection of the carcinogen over the life span of the erythrocyte, which lasts up to four months in humans [26]. Levels of hemoglobin adduct of 4-aminobiphenyl (4-ABP), which is a potent human urinary bladder carcinogen and is present in significant quantities in tobacco smoke [27], has been shown to accurately and quantitatively reflect levels of exposure in both smokers and nonsmokers [27]. Adduct levels also decline in a predictable manner upon cessation of smoking. Also, in studies of smokers using different types of tobacco, adduct levels clearly reflected metabolic capability for metabolizing the compound in a manner that also correlated with elevated risk of bladder cancer [28]. Further, studies of newborn infants of smoking and nonsmoking mothers revealed the presence of 4-aminobiphenyl adducts in hemoglobin of the infants at levels reflective of the exposure history of the mothers. Thus, this approach provides a sensitive and accurate means of monitoring exposure to maternal metabolites and demonstrates the technique's potential to monitor exposures to a wide variety of carcinogens [29]. Elevated hemoglobin adducts have been observed after Alkenes and aromatic amines exposures [30].

Another protein is serum albumin, which because of a relatively long half-life time (2-3 months) was shown to be a reliable adduct carrier for Aflatoxin B_1 (AFB_1) the potent liver group I human carcinogen [5] associated with increased incidence of hepatocellular carcinoma in Africa and Southeast Asia, especially in the presence of hepatitis infection [31]. AFB_1 has been detected in peripheral blood as AFB_1-albumin adduct, which is considered as the most reliable biomarker of AFB_1 exposure in humans [32] and has been shown to be related to hepatocellular carcinoma risk in a dose-response fashion among HBV surface antigen (HBsAg) carriers [33].

DNA-Adducts of carcinogens are usually mutagens and are thought to initiate tumorigenesis by reacting with DNA, which can be misreplicated when encountered by a DNA polymerase [25, 34]. Adduct measurement constitutes not only a means of identifying the early lesions in the development of cancer in situations in which, for statistical reasons, the risk is too low to be detectable, it also gives a possibility of identification of the dangerous components in mixed exposures and of the relevant reactive intermediates in cases of complex metabolism, However, it is difficult to determine definitively which type of adduct caused which mutation in multiple exposures [35]. In liver biopsies from both human and animal liver samples, ultra-sensitive and highly specific methods have been developed for the identification and quantitation of the N^2, 3-ethenoguanine (EG) DNA adducts associated with exposure to vinyl chloride, which is one of the well known human carcinogen [36, 37]. Ultraviolet light induces two main types of photoproduct in cellular DNA, cyclobutane pyrimidine dimmer and 6-4 photoproduct. The persistence of these lesions can interfere with essential processes such as transcription and DNA replication, possibly leading to cell death, mutation or neoplastic transformation. Quantitation of these photoproducts provide more meaningful information on exposure and, ultimately, on risk of cancer development. [38]. The role of AFB_1 exposure in human cancer incidence has been elucidated by measuring levels of the covalent adducts of AFB_1 on DNA quickly and effectively and comparing its dose with the effects of both hepatitis B and C among patients with hepatocellular carcinoma [39]. Measuring Benzo[a]pyrene-diol-epoxide-DNA adducts by immunoaffinity techniques has been shown to be a good means of monitoring human exposure to the ubiquitous environmental pollutant polycyclic aromatic hydrocarbons, which arises from incomplete combustion of organic materials [40]. Also, occupational exposure to the carcinogen styrene, a volatile monomer widely used in the production of polymers and reinforced plastics, results in the formation of styrene oxide-DNA adduct in circulating human mononuclear cells [41].

1.3. Xenobiotic Itself and/or Specific Metabolites

Traditional definition of an exposure biomarker involves measurement of a xenobiotic or its metabolite in a body tissue or fluid, bearing into consideration the factors that influence the choice of tissue or fluid and analyte, which include the toxicodynamics of the agent (where it is absorbed, distributed, assimilated stored, excreted), speciation (its chemical form), and chronology (acute vs. chronic exposures and temporal aspects of its kinetics in the body) [42]. In some instances an exposure marker can distinguish long-term from short-term exposures, where blood lead may reflect recent exposure, while bone lead reflects cumulative exposure [43]. Blood lead is a marker of exposure and a surrogate marker of effect. Elevations of erythrocyte protoporphyrin and urinary α-aminolevulinic acid (α -ALA) or the activity of α -ALA dehydrates are all effects caused by lead; the inhibition of enzymes in the heme synthesis pathway results in anemia, yet anemia is rarely the end point of concern in lead poisoning, especially in adults. Thus, they are useful as exposure as well as effect markers. [43].

Cotinine, a metabolite of nicotine, measured in blood, saliva, or urine appears to be a reliable biomarker of recent smoking status in population whether active and passive smokers [44].

2. Biomarkers of Effects

The International Programme on Chemical Safety has defined a biomarker of effect as "A measurable biochemical, physiological, behavioral or other alteration within an organism that, depending upon the magnitude, can be recognized as associated with an established or possible health impairment or disease" [20]. Biomarkers of effect are on the borderline between healthy surveillance, even diagnostics and monitoring of exposed populations. They can be endogenous induced components, or a measure of the functional capacity or some other indicators of the state or balance of the body organ system, as affected by exposure. Such effects markers are generally preclinical indicators of abnormality and may be specific or non-specific. At the molecular level biomarkers of effect can be categorized into biomarkers of genotoxic effects [45] and biomarkers of response [46].

2.1. Biomarkers of Response

In multicellular organisms, general biological principles postulate that the organization of cells depends on regulated cell surface interactions both with molecules (surface ligands) on the surface of other cells (particularly immune cells) and with immune mediators in the extracellular fluid [47]. Concurrently, these extracellular ligands trigger nuclear signals through a cascade of intercellular protein-protein interactions that are known to generate and/or require reactive free radicals or derived species to successfully transmit their signals to the nucleus [47, 48]. Most environmental agents are reactive oxygen species inducing agents [49, 50] and consequently sub-toxic exposures of these free radical inducing agents can cause altered expression (increase or decrease) of proteins involved in multiple signal transduction pathways. Also, there are inducible proteins whose genes are highly influenced by environmental stimuli. Some of these proteins represent protective mechanisms against different environmental stresses [19, 51, 52], while others amplify adaptation to environmental conditions [53, 54].

Response at the molecular level can be categorized into: a) activation of Nuclear Transcription Factors, b) activation of oxidant enzymes, c) activation of signal transduction proteins involved in growth pathway, d) interference with gap junction, e) activation of adaptive response.

2.2. Activation of Nuclear Transcription Factors

Transcription factors are DNA-binding proteins that regulate gene expression of a variety of genes [55], including DNA-repair genes [56] cytokines and growth factors, adhesion molecules, immunoreceptors and other biological activities related to DNA response to external stimuli [51] The activity of transcription factors is typically regulated by phosphorylation-dependent events that can include the phosphorylation of the transcription factor itself, which is then followed by its translocation to the nucleus [57]. Transcription factor proteins main functions are to enhance the transcription phase proteins and are required for maximal transcription of many cytokines, including tumour necrosis factor-(TNF-), interleukin-1 (IL-1), IL-6, and IL-8, which are thought to be important in the generation of acute inflammatory responses [58, 59]. T cell activation results in a rapid activation of transcription factors, which appear to play an important role in T cell proliferation, partly because of the involvement of these transcription factors in the up-regulation of the IL-2 and cognate receptor (IL-2R) genes [60, 61].

An example of these nuclear transcription factors is the nuclear factor-κB (NF-κB). It was originally described to be composed of heterodimeric subunits of p50 and p65 proteins [62, 63]. It is typically found in the cytoplasm of cells as an inactive precursor complexed with inhibitor proteins [64]. Following activation by external stimuli, the inhibitory units are phosphorylated and degraded, unmasking nuclear localization signals that allow NF-κB to be transported to the cell nucleus, where its dimers are free to bind DNA containing the sequence (5'- GGGRNNYYCC -3'), where R is purine, Y is pyrimidine, and N is any base [65, 66]. Many of the NF-κB-targeting genes are pivotal in mediating cell-to-cell interaction, intercellular communication, cell recruitment or transmigration, amplification or spreading of primary pathogenic signals [67]. NF-κB is required for maximal transcription of many cytokines, including tumour necrosis factor-(TNF-), interleukin-1 (IL-1), IL-2, IL-6, and IL-8, which are thought to be important in the generation of acute inflammatory responses [65 68]. NF-κB activation, was shown to be stimulated by bacterial endotoxin, chemical mitogens, viral proteins [9], UV light [69], and certain chemical agents including ozone [70] arsenic and chromium [71, 72] and asbestos [73]. Excessive activation of NF-κB in leucocytes is stimulated by ionising radiation in a dose response pattern [74]. There is an increasing body of evidence suggesting a role for NF-κB in carcinogenesis [63]. For example, NF-κB is implicated in signaling tumor promoter-induced transformation and is activated by viral transforming proteins [75-76]. The importance of NF-κB cannot be overstated, as failure in any of the mechanisms leading to NF-κB activation can have serious consequences for the cell [63].

Studies involving NF-κB are frequently compared with those involving activator protein-1 (AP-1) [61]. AP-1 is a family of accessory transcription factors that interact with other regulatory DNA sequences. The AP-1 binding site is TGAGTCA (called TPA response element, TRE or AP-1 sites) [77, 78]. The family of genes that interact with AP-1 sites includes both homo- (Jun/Jun) and heterodimeric (Fos/Jun) complexes encoded by various members of the c *-fos* and c *-jun* families of protooncogenes. There are three Jun proteins (i.e. c-Jun, Jun B, and Jun D) and four Fos proteins (i.e. c-Fos, Fos B, Fra-1, and Fra-2). AP-1 binds to the TRE in the promoters of AP-1-induced genes and mediates transcriptional activation of these genes [77, 79]. Both c-*fos* and c-*jun* are immediate early response genes associated with transition of the cells from the G1 stage of the cell cycle to *S* phase and their induction may provide the molecular switch for programmed cell death

[80], cell proliferation [81], or transformation of cells [82, 83]. The magnitude of AP-1 activation was suggested to predict transformation response and tumor promotion [81-83]. Oxidative stress [79], asbestos [73, 84], DDT [85], ionizing radiation [86, 87] and both arsenite and arsenate [88, 89] are examples of environmental agents, shown to induce transactivation of AP-1.

Another example of nuclear transcription factors is metal transcription factor-1 (MTF-1), which is an evolutionarily conserved protein that regulates the transcription of the inducible protein metallothioneins (MTs) [90, 91]. MTs are a family of evolutionarily conserved, low molecular- weight, cysteine-rich, metal-binding proteins [92, 93], whose proposed roles include *a*) participation in maintaining the homeostasis of essential transition metals, *b*) detoxification of nonessential metals, and *c*) protection against intracellular oxidative stress [92-95]. MTs transcription is mediated by regulatory elements called metal-responsive elements (MREs). MREs contain a 7-bp core sequence (TGCRCNC) and are present in multiple copies in the promoter/enhancer regions of almost all metal-inducible MTs [96, 97]. MTF-1 specifically binds to MREs and phosphorylation is involved in its activation in response to metal ions [98] and has been demonstrated to contain six zinc finger domains and several *trans*-activation domains: acidic-rich, proline-rich, and serine/threonine-rich, all of which are required for metal-inducible transcription [99]. Ionizing radiation [94], zinc, cadmium [98], viral infection [100], hyperoxia [101] and hypoxia [102] are examples of environmental agents that activate MTF-1.

2.3. Activation of oxidant enzymes

Hydrogen peroxide (H_2O_2), Hydroxyl radical ($\cdot OH$), superoxide (O_2) and nitric oxide (NO) radical are considered a nominal set of biologically relevant free radicals, which are released during oxidative metabolism under normal physiologic conditions [103]. They are formed as intermediates in mitochondrial electron transport systems and microsomal metabolism of endogenous compounds and xenobiotics, including drugs and environmental pollutants and various cytoplasmic sources [51]. Numerous various enzymes including: NADPH oxidase [104], xanthine oxidase [105] and nitric oxide synthase [106] generate them. In addition, neutrophils and other inflammatory cells generate and release reactive oxygen species via an NADPH-oxidase-dependent mechanism that is mediated by membrane receptor activation of protein kinase C and phospholipase C [51, 52]. One of the major functions of these free radicals is immunological host defense, where they are generated by macrophages and neutrophils and play critical roles as bactericidal, antiviral and anti-tumor agents [51, 52, 106]. The increase in production of ROS following various external stimuli has been associated with various cellular pathways of gene expression [107] such as cell proliferation [108] or apoptosis [109], where these free radicals are designated to play a significant role in the molecular response to cellular stress by directly activating gene transcription [110] and signal transduction [48]. Generation of such reactive oxygen species to a level that overwhelms tissue antioxidant defense systems, results in an oxidative stress whose magnitude depends on the ability of the tissues to detoxify such free radicals [111], and consequently damaging cellular lipids, proteins and DNA inducing lipid peroxides, protein carbonyls and DNA damage [49, 50, 112]. Environmental agents which generate free radical are numerous [49-52] and include various food sources [113], infectious organisms [10, 106], most physical and chemical agents including ionizing radiation [114], asbestos [115], and are also provoked during exercise [116] and most diseased conditions [50, 117, 118].

Nitric oxide is a biologically active molecule produced in several types of mammalian cells including PMNs, macrophages and muscle cells [119]. It is a fascinating free radical molecule which participates in a broad range of important physiologic processes, including vasodilatation, neurotransmission, and host defense [119]. NO is generated from the amino acid l-arginine by three isoforms of the enzyme NO synthase; the constitutive (cNOS), the endothelial (eNOS) and the inducible (iNOS). The inducible form generates much larger amounts of NO (1000 times fold than the other two isoforms and its cellular production continues for many hours. Accordingly, iNOS effects are much better spread [120]. Inducible NOS has been detected in virtually every cell type, and the NO that it produces can perform both beneficial and detrimental actions, where, excess production of NO can cause tissue damage and contribute to the development of a wide spectrum of diseases including septic shock, rheumatoid arthritis, cerebral ischemia, multiple sclerosis, and diabetes [121]. However, physiological amounts can eliminate infiltrating microorganisms (106), reduce thrombosis, and improve blood supply to injured tissues, as it is the key signal molecule in cell-cell interactions controlling vascular regulation [122] and neuronal communication [123]. Recent findings indicate that iNOS represent an important repair mechanism in response to vascular injury [124]. Similar investigations demonstrated that metabolic evidence for iNOS is detectable in wounds during the initial 48 - 72 hours in response to repair processes and that tissue macrophages express NOS during early phases of tissue repair [125, 126]. Agents known to induce the expression of iNOS mRNA are numerous and some of them include UV [127], ionizing radiation [128] Helium neon laser [129] and asbestos [130].

2.4. Activation of Growth signal Transduction Proteins

Many cancers and some other environmentally related diseases are believed to be associated with the aberrant expression of genes that encode proteins involved in cellular growth signal transduction, including growth factors and the protein products of oncogenes and tumor-suppressor genes involved in the regulation of cellular proliferation [19, 131-137]. These aberrant expressions, which are easily accessible in extracellular fluids, can involve a quantitative difference from normal (i.e., overexpression of the wild-type protein) and/or exhibit a qualitative difference, when expresses the mutant forms of the protein [138]. Growth signal transduction requires the concerted actions of several signaling molecules including growth factors, growth factor receptors, G-proteins, protein kinases, oncoproteins, suppressive genes and cyclins.

a) Increased expression of cell cycle positive extracellular controls

Growth factors are usually small proteins that are secreted into the extracellular to provide intercellular communication that regulates cellular proliferation. They are increased in expression following damage induced by external stimuli to any tissue, where repair of injury begins as soon as tissue damage occurs, and the release of peptide growth factors from injured cells and inflammatory cells is a critical part of this repair process [138]. Growth factors probably also play important roles in spontaneous and chemically induced cancer and other non-cancer proliferative diseases [139].

Epidermal growth factor (EGF) is elevated in the serum of some patients with stomach cancer [140], cancer of the tongue [141], ovarian cancer [142] and has been shown to be elevated among pyrethroids pesticide exposed workers [143] and due to asbestos exposure [144]. Insulinlike growth factors (IGF) have been reported to be elevated in the plasma of some breast [149] and ovarian cancer patients [145], and is observed in the blood of normal subjects after alcohol ingestion [146]. Elevated serum levels of platelet derived growth factor (PDGF) have been implicated in the development of pulmonary fibrosis induced by asbestos in a model system [147]. Transforming growth factor α (TFG-α) has been shown to be involved in arsenic-mediated skin neoplasia [148] and is expressed at maximum levels post exposure to low doses of ionizing radiation [149]. Hepatocyte growth factor (HGF) have been reported in hepatocellular carcinoma as well as in nonmalignant liver diseases [150]. In addition, basic fibroblast growth factor bFGF have been identified in biological fluids of patients of a wide spectrum of cancer diseases [136] and in serum of pesticides exposed workers [143] and asbestos exposed workers [151].

Transmembrane growth factor receptors are proteins, which consist of an extracellular binding domain involved in signal transduction from an extracellular growth factor across the cell membrane, a transmembrane domain and an intracellular enzymatic domain. Such proteins are frequently over-expressed in human malignancies. Overexpression is accompanied in many instances by cleavage of the extracellular domain (ECD) of the receptor with its accumulation in the extracellular environment. Thus, detection of increased amounts of the ECD of these receptors in blood is a potential biomarker of cancer development [152].

The ECD of the epidermal growth factor receptor EGFR, encoded by the *cerbB*-1 oncogene) has also been identified in cancer patients and has been identified as an early biomarker of malignant disease [152]. EGFR has been shown to be significantly elevated in lung cancer patients with asbestosis compared to asbestosis controls without cancer matched for age, sex, race, smoking, and asbestos exposure, and 20 nonasbestosis, non-cancer controls matched for age, sex, race, and smoking [152, 153]. Serum EGFR ECD levels, has also been found to be statistically significantly elevated among 22 former uranium miners with lung cancer compared to healthy controls [135]. The ECD of EGFR has also been measured in the urine of squamous cell carcinoma patients. Elevated levels of EGFR were detected in 15 of 42 (36%) squamous cell carcinoma patients compared to 8 of 50 (16%) non-squamous cell carcinoma patients and 3 of 50 (6%) non-cancer controls, ($p<03$) [154].

Decreased expression of EGFR mRNA post exposure to dioxin has been identified in rat uterine model [155]. Similarly humans with low levels of blood dioxin exposure show significant higher levels of EGFR ECD compared to individuals with medium to high blood dioxin levels [156].

b) Increased expression of cell cycle positive intracellular controls

Cyclins and cyclin-dependent protein kinase (CDK) Complexes are the engines that drive progression of cell cycle from one-checkpoint functions of G_1, S, and G_2 phase to the next. These are a series of protein complexes composed of two subunits cyclins and cyclin-dependent protein kinases (CDK). Among G_1 and G_2 checkpoints regulators is the D-type cyclins, which serve as targets of growth factors to integrate extracellular signals into the core cell cycle regulators [157]. Much evidence has accumulated to prove that one of the most important functions of cyclin D/Cdk4 complex is to phosphorylate retinoblastoma protein (pRb) in mid to late G1 and to neutralize the anti-proliferative activity of retinoblastoma suppressor gene product (pRb)'[158]. D-type cyclins are composed of three different but closely related subfamilies (D1, D2, and D3), all differentially expressed in a wide variety of organs and in a tissue-specific manner [159]. D-type cyclins are induced to express in response to a variety of mitogenic signals and function as a regulatory subunit of cyclin-dependent kinases (Cdk) [160]. D-type cyclins can interact with 4 different Cdks (Cdk2, 4, 5, and 6), among which Cdk4 and Cdk6 are apparently the major functional catalytic partners in proliferating cells [160, 161]. When cells are exposed to growth factor stimulation, the expression of cyclin D is maintained regardless of the point in the cell cycle [159]. However, accumulation of active cyclin D/Cdk4 (or Cdk6) complex is rate-limiting and is required for cells to progress through G1 and to commit to entering S phase [162]. The overexpression of D-type cyclins shortens the length of G1 without affecting remainder of the cell cycle [163]. Demonstrating the impact of nongenotoxic chemicals on cell cycle checkpoint function showed that agents such as ionizing radiation [164], asbestos [165, 166], and UV [167] induce cell cycle disruptions at G1 and G2 phases. These alterations in cell cycle progression occur after exposure to DNA damaging agents. This is likely to permit DNA repair before the continuing of DNA synthesis or mitosis. Delays take place at both G1 and G2 phases of the cell cycle [168]. Failure in repair processes could result in fixation and propagation of genomic changes necessary for neoplastic transformation [169]. Under certain circumstances, DNA damage may trigger a rapid programmed death of the cells, apoptosis [168].

c) Increased expression of signal transduction proteins

G proteins (guanine nucleotide binding) are proteins encoded by the H-*ras*-1 gene, which is protooncogene responsible for control of cell growth (proliferation) and specialization (differentiation) and is implicated in a variety of tumor types. This gene exerts a powerful proliferative response through signal transduction [170], which is an integrated cascade of multiprocess system responsible for the smooth, orderly, and specific transmission of extracellular signals to the nucleus. The gene proteins ras-p21 are membrane bound with GTPase activity, and form complexes with other proteins that exert a powerful proliferative response through signal transduction [171]. Ras oncogenes can be activated by the introduction of specific point mutations at codons 12, 13, 61 and 119 due to exposure to vinyl chloride [172], 1,3-butadiene [173] and ionizing radiation [174]. Ras oncogene can be also activated by over expression of the normal proto-oncogene [133] and are found to be elevated in serum samples from pneumoconiosis (asbestosis and silicosis), where ras-p21-positive patients had elevated levels years prior to the clinical detection of disease in a variety of tumor types [133]. The activation of ras oncogene and consequent expression of G-proteins is a consequent of exposure to DNA damaging agents [175] with and is activated due to exposure to UV [176] and viral infection [177].

d) Increased expression of cell cycle negative extracellular controls

There are certain proteins that are known to inhibit cell proliferation. The retinoblastoma suppressor gene product (pRb) activates Transforming growth factor beta (TGF-ß) gene expression by stimulating nuclear proteins that bind to its promoter [178]. TGF-ß is a multifunctional cytokine that enhances extracellular matrix formation, inhibits cellular proliferation, and suppresses immune processes [179, 180], by regulating immune cell function, where TGF-ß1 reduces local production of growth factors and tissue damage induced by free radicals [181]. It is believed that TGF-ß1 exerts both positive and negative influences on cancer development [182]. TGF-ß1 promotes remodeling of the extracellular matrix, which may mediate tumor cell-matrix interactions and epithelial cell differentiation (183-185). Several external insults can stimulate the production of TGF-ß. These include asbestos [186], ionizing radiation [187], infrared [188] and UV [189]. Nickel-induced transformation is associated with abrogation of TGFß growth control [190].

e) Increased expression of cell cycle negative intracellular controls

Several families of genes encode for proteins playing important roles in negative growth control, including growth suppression. These include the GADD45 family (GADD45alpha, beta, and gamma) [191], The retinoblastoma gene family (pRb) [192], p21Wafl protein, which is a universal inhibitor of cyclin-dependent kinases (CDKs) [193] and gene p53 [194]. The latter, being extensively studied, is an ideal biomarker of effect for addressing such questions of exposure and risk. The p53 gene is expressed in every cell of the human body and the p53 protein has been shown to play a crucial role in the regulation of cell division, making it a central molecule in the life of a cell [195]. The p53 protein can influence the cell cycle in several ways. It can cause G_1 and G_2 growth arrest or apoptosis [196, 197]. It is mutated in the majority of human tumors, which indicates that its role is important for prevention of malignant transformation [198, 199]. Inactivation of the p53 gene can result in uncontrolled cell division and ultimately lead to cancer [199]. Such inactivation frequently occurs as a result of mutations caused by exposure to environmental mutagens. However, inactivation of p53 can also occur sporadically through endogenous processes and, in certain cases, mutations in p53 can be inherited [200]. As a biomarker of effect, elevated serum levels of total p53 protein have been reported in workers with chromium exposure compared to non-exposed controls [201]. Similar results are observed among uranium miners [137], vinyl chloride plant workers [202] and is associated with exposure to dietary aflatoxin B_1 (AFB_1) leading to hepatocellular carcinoma (HCC) of the liver [203].

f) Disruption of cell-cell communications

Gap junctional intercellular communication (GJIC) is considered to play a key role in the maintenance of tissue independence and homeostasis in multicellular organisms by controlling the growth of GJIC-connected cells, since extracellular, intracellular, and intercellular communication crucially participate in the epigenetic changes of gene expression [204-205]. Intercellular communication through gap junction links a cell with its neighbors in a tissue, and thus individual cells exist communally as a multicellular organism [205-208]. Development of tumors results in uncontrolled cellular growth due to down regulation of GJIC. Accordingly, most cancer cells do not behave in harmony with their neighbors [204, 207]. Evidence supports the hypothesis that inhibited GJIC activity is related to carcinogenesis [207, 208]. Gap junction channels are composed of connexin molecules and, so far, more than a dozen different connexin genes are expressed in mammals [209]. Evidence is accumulating that connexin (Cx) genes form a family of tumor-suppressor genes [210] and evidences suggest that mutations of connexin genes can affect the tumor-suppressive function of gap junction and that gap junctional intercellular communication can be regulated by both genotoxic and non-genotoxic environmental carcinogens [210]. Of these agents ionizing radiation [211], polycyclic aromatic hydrocarbons [212], pesticides [213], UV [214], electromagnetic waves [215], polyethylene [216] and cigarette smoke [217].

g) Activation of stress (adaptive) response

It has been long established that when organisms or cells are exposed to low levels of specific harmful physical or chemical agents, a beneficial physiologic effect is observed. Concurrently, exposure to sublethal challenges of stress may rejuvenate the cell by repairing damage before the challenge and may provide transient protection against further damage from subsequent sublethal or lethal challenges with a different otherwise harmful physical or chemical stressor [53, 54, 218]. It has been suggested that a mechanism contributing to these beneficial effects of otherwise harmful agents may operate through the stress response (SR) [53, 132, 219], a physiologic mechanism characterized by the induction of families of detoxification and repair molecules (heat shock proteins) [220], induction of DNA repair and antioxidant enzymes [221] and programmed cell death [222]. The SR has been characterized using heat (the classical SR), radiation, heavy metals, and oxidizing agents [53, 54].

Antioxidant Enzymes encoded by numerous genes in mammalian systems have been shown to be responsive to oxidants, although a systematic mechanism for gene regulation by oxidative stress has not been elucidated. Oxidative stress has been shown to alter the expression of mammalian antioxidant enzymes including superoxide dismutase, glutathione peroxidase (GPx), ά-glutamylcysteine synthetase, catalase, glutathione S-transferase and quinone reductase [223-224]. Induction of haem oxygenase by oxidative stress may both function in intracellular signalling and serve to protect cells from further oxidant injury [225]. Antioxidant enzymes especially the inducible superoxide dismutase enzyme has been shown to be elevated in individuals at risk of exposure to low doses of various stresses. Cigarette smokers [226], asbestos exposed workers [227], ionizing radiation exposed workers [228] and athletes [229] have higher levels of antioxidant enzymes compared to controls. Also the antioxidant enzyme superoxide dismutase in samples from patients with malignant tumors revealed high levels of superoxide dismutase protein, up to 45-fold greater than the level of controls [230, 231]. This increased level seems to be the cause of decreased apoptotic response in tumour cells [232, 233].

Stress proteins are polypeptides encoded by genes called the stress genes. When cells are exposed to heat or stress, abnormal protein folding directs the increased expression of stress genes to induce polyubi-quitin [234] and other heat shock proteins [235].

Heatshock proteins are constitutively expressed at high levels in all cell types under normal conditions (5-10 % of the total protein content). They have anti-apoptotic effects [236] as they are cytoprotective and essential to a variety of signal transduction pathways, and are proposed to recognize hydrophobic regions that are expressed on the surface of the misfolded proteins, facilitate their refolding and thus prevent their reversible aggregation [237, 238]. Scientific evidence indicates a low degree of heat shock protein expression can be detected in the well-differentiated carcinomatous areas, but not in the poorly differentiated areas [239, 240]. Also, esophageal squamous cell carcinoma with HSP70 expression exhibited a significantly better prognosis compared with HSP70-negative esophageal squamous cell carcinoma in a univariate analysis [241]. It was suggested that these proteins have independent functional roles for an anti-tumoural response [240, 241]. Exposure to several environmental agents UV [240], ionizing radiation [242], cadmium [243] and heat [235] induces the production of heat shock proteins.

Two classes of *ubiquitin genes* have been identified in eukaryotes: ubiquitin fusion genes, which encode a single ubiquitin moiety fused to an unrelated ribosomal tail protein [234]; and polyubiquitin genes, which consist of head-to-tail tandem repeats of complete or partial ubiquitin coding units [244, 246]. The ubiquitin fusion genes are expressed constitutively in the cell, whereas the polyubiquitin genes are induced in the stress response or transcribed under specific physiologic conditions [246-247]. A regulatory role for ubiquitin is suggested by the reversible covalent modification by ubiquitination of substrates such as calmodulin, histones H2A and H2B, as well as several cell-membrane receptors [244-247] and by the fact that ubiquitin is required in a wide variety of cellular functions such as cell-cycle control, DNA repair, DNA replication, differentiation, peroxisome biogenesis, and heavy metal resistance [234]. Ubiquitin genes are expressed during muscular exercise [248], ionizing radiation [249] and UV [250].

Induction of Apoptosis: The homeostatic balance between cell proliferation and apoptosis in the maintenance of constant cell numbers may provide a hormetic effect by minimizing the consequences of proliferation-related mutagenesis during tumor promotion [251-253]. The adaptive increase in apoptosis that accompanies the oncogene-activated dysregulation in proliferation selectively eliminates potentially preneoplastic cells in hyperplastic foci. Acquired resistance to apoptosis appears to be a pivotal event in the transition to malignancy [255]. Apoptosis is induced by various environmental agents which are known to elicit an oxidative stress [109, 233], including: arsenite [88] ionizing radiation [242],], exercise [255] and cigarette smoking [256].

DNA repair involved genes are critical for maintaining the integrity of genetic material transmitted from one cell to another and for protection against mutations leading to cancer. Mammalian cells rely on DNA repair systems to maintain their genomic integrity. DNA-repair is expressed in several forms; the base excision repair, nucleotide excision repair, mismatches repair, recombination and rejoining pathways [257]. Recent evidence suggests that the phenomenon of mutator phenotype leading to genetic instability may be caused by DNA repair deficiency [258, 259]; this repair deficiency may be caused by exposure to environmental toxicants. For example, tire-storage workers exposed to nitroso-compounds and hospital personnel exposed to antineoplastic drugs were deficient in methylguanine alkyltransferase activities [260, 261]. Abnormal repair synthesis was documented in lymphocytes of cigarette smokers and drug addicts when their cells were challenged with UV light or chemicals [262- 264]. In general cells possess an efficient DNA repair system from which only a few damaged DNA lesions can escape [50]. Increased susceptibility to breast and lung cancers and hematologic malignancies are characterized by reduced DNA repair capacity [265, 266]. The development of leukemia and lymphoma in cases of Fanconi anemia or ataxia telangiectasia is associated with defective DNA repair [267, 268].

3. Biomarkers of Susceptibility

Biomarkers of susceptibility measure innate or induced individual capabilities for response to exposure to environmental toxicants [20]. Potential sources of susceptibility for cancer risk include individual variation in DNA repair capacity for specific types of DNA damage [265-270]. Also individual variation in enzymes which activate and detoxify procarcinogens and carcinogens (e.g., Phase I enzymes, which catalyze oxidation, reduction, and hydrolysis reactions, and Phase II enzymes, which catalyze conjugation and synthetic reactions) are causes for diffferential susceptibility. Functional polymorphism have been identified primarily at enzymes associated with redox regulation and detoxification, such as glutathione *S*-transferase [271, 272] and cytochrome p450 isozymes [273, 274]. Cancer susceptibility can be the inability to eliminate mutated cells by apoptosis due to mutation in suppressive genes [197], and/or induced disruption in gap junction [213], activation of protooncogene [275] and/or inactivation of suppressive genes [276].

Inherited susceptibility due to a defective gene is a factor in a small percentage of people who develop cancer (less than 5%), while *induced susceptibility* is due to the wide variation in individual responses to exogenous agents is believed to result from the great diversity in responsiveness to risk factors in the environment. These variations, known as polymorphisms, are caused by sporadic mutations caused by both endogenous and exogenous processes [20]. In most instances, such mutations result in minor changes in the nucleotide sequence of the coding region as well as 5' and 3' untranslated regions, which are sufficient to alter expression or stability at both the RNA and protein levels [277]. However, there are many instances where modifications in gene expression do not involve changes in DNA nucleotide sequences. Modifications in gene expression through methylation of DNA and remodelling of chromatin via histone proteins are believed to be the most important events of the epigenetic changes [278].

4. Conclusion

Detectable gene mutations and alterations in signal transduction pathways together with modified post-translational proteins offer valuable molecular biomarkers for occupational and environmental human biomonitoring applied for the identification of potentially hazardous exposures before adverse health effects appear and allow the establishment of exposure limits in order to minimize the likelihood of significant health risks. Concurrently, most of these elicited gene expressions are also expressed in tissue transformation and progression of tumors and are similar to the hallmarks used for cancer prognosis. Some of these proteins represent protective mechanisms against different environmental stresses, while others amplify adaptation to environmental conditions. The resultant balance between protective proteins and adaptive proteins seem to determine an individual likelihood to develop cancer.

References

[1] B. Helmut and E. Hietanen, The Role of Individual Susceptibility in Cancer Burden Related to Environmental Exposure, *Environ. Health Perspect.* 104(Suppl 3) (1996) 569-577.
[2] M. Kraine and R. Tisch, The Role of Environmental Factors in Insulin-Dependent Diabetes Mellitus: An Unresolved Issue, *Environ. Health Perspect.* 107(suppl 5) (1999) 777-781.
[3] G. Joshi *et al.*, Iron and Aluminum Homeostasis in Neural Disorders, *Environ. Health Perspect.* 102(Suppl 3) (1994) 207-213.
[4] R. Doll and R. Peto, Cigarette Smoking and Bronchial Carcinoma: Dose and Time Relationships among Regular Smokers and Lifelong Non-smokers, *J. Epidemiol. Community Health* 32 (1978) 303-313.
[5] S. Henry *et al.*, Policy Forum: Public Health. Reducing Liver Cancer--Global Control of Aflatoxin, *Science* 286(5449) (1999) 2453-4.
[6] B. Mossman and J. Gee, Medical Progress. Asbestos-related Diseases, *N. Engl. J. Med.* 320 (1989) 1721-1730.
[7] F. Urbach, Ultraviolet Radiation and Skin Cancer of Humans, *J. Photochem. Photobiol. B.* 40 (1997) 3-7.
[8] H. Nikjoo *et al.*, Modeling of Radiation-induced DNA Damage: The Early Physical and Chemical Event, *Int. J. Radiat. Biol.* 66 (1994) 453-457.
[9] J. Parsonnet, Bacterial Infection as a Cause of Cancer, *Environ. Health Perspect.* 103(Suppl 8) (1995) 263-268.
[10] W. Spannhake *et al.*, Synergism Between Rhinovirus Infection and Oxidant Pollutant Exposure Enhances Airway Epithelial Cell Cytokine Production, *Environ. Health Perspect.* 110 (2002) 665-670.
[11] A. Badawi, Molecular and Genetic Events in Schistosomiasis-associated Human Bladder Cancer: Role of Oncogenes and Tumor Suppressor Genes, *Cancer Lett.* 105(2) (1996) 123-38.
[12] A. Hartmann *et al.*, Does Physical Activity Induce DNA Damage? *Mutagenesis* 9 (1994) 269-272.
[13] R. Frentzel-Beyme, Depression as a Risk Factor for Cancer: Re-newing a Debate on the Psychobiology of Disease, *J. Natl. Cancer Inst.* 91 (1999) 1586–7.

[14] W. Willett *et al.*, Dietary Fat and Fiber in Relation the of Breast Cancer: An Eight Year Follow Up, *JAMA* 268 (1992) 2037-2044.

[15] W. Xing and R. Na, Amino Acids in Excess Increase SCEs in Human Lymphocytes, *Mut. Res.* 372 (1996) 75-78.

[16] L. Martin *et al.*, The Role of Reactive Oxygen and Nitrogen Species in Airway Epithelial Gene Expression, *Environ. Health Perspect.* 106(Suppl 5) (1998) 1197-1203.

[17] J. Hamilton and K. Wetterhahn, Differential Effects of Chromium (VI) on Constitutive and Inducible Gene Expression in Chick Embryo Liver *In Vivo* and Correlation with Chromium (VI)-induced DNA Damage, *Carcinogenesis* 2 (1989) 274-286.

[18] J. Hamilton *et al.*, Genotoxic Chemical Carcinogens Target Inducible Genes *In Vivo*, *Ann. NY. Acad. Sci.* 726 (1994) 343-345.

[19] N. Johnson *et al.*, DNA Damage-inducible Genes as Biomarkers for Exposures to Environmental Agents, *Environ. Health Perspect.* 105(Suppl 4) (1997) 913-918.

[20] J. Barrett *et al.*, 12th Meeting of the Scientific Group on Methodologies for the Safety Evaluation of Chemicals: Susceptibility to Environmental Hazards, *Environ. Health Perspect.* 105(Suppl 4) (1997) 699-737.

[21] A. William *et al.*, Human Biomonitoring: Research Goals and Needs, *Environ. Health Perspect.* 104(Suppl 3) (1996) 479-483.

[22] D. Christiani, Utilization of Biomarker Data for Clinical and Environmental Intervention, *Environ. Health Perspect.* 104(Suppl 5) (1996) 921-925.

[23] M. Andersen and H. Barton, The Use of Biochemical and Molecular Parameters to Estimate Dose-Response Relationships at Low Levels of Exposure, *Environ. Health Perspect.* 106(Suppl 1) (1998) 349-355.

[24] N. Rothman and R. Hayes, Using Biomarkers of Genetic Susceptibility to Enhance the Study of Cancer Etiology, *Environ. Health Perspect.* 103(Suppl 8) (1995) 291-5. Review.

[25] L. Ehrenberg *et al.*, Macromolecule Adducts as Biomarkers of Exposure to Environmental Mutagens in Human Populations, *Environ. Health Perspect.* 104(Suppl 3) (1996) 423-428.

[26] M. Meyer and W. Bechtold, Protein Adduct Biomarkers: State of the Art, *Environ. Health Perspect.* 104(Suppl 5) (1996) 879-882.

[27] C. Patrianakos and D. Hoffmann, Chemical Studies of Tobacco Smoke. On the Analysis of Aromatic Amines in Cigarette Smoke. *J Anal. Toxicol.* 3:150-154 (1979).

[28] P. Santo *et al.*, Levels of the Adducts of 4-aminobiphenyl to Hemoglobin in Control Subjects and Bladder Carcinoma Patients, *Cancer Lett.* 60 (1991) 245-251.

[29] G. Wogan, Molecular Epidemiology in Cancer Risk Assessment and Prevention: Recent Progress and Avenues for Future Research, *Environ. Health Perspect.* 98 (1992) 167-178.

[30] S. Tannenbaum, and P. Skipper, Blood Protein as Carcinogen Dosimeters, Basic Life Sci. 38 (1986) 473- 478.

[31] J. Groopman *et al.*, Molecular Dosimetry of Aflatoxin-N7-guanine in Human Urine Obtained in The Gambia, West Africa, *Cancer Epidemiol. Biomarkers Prev.* 1 (1992) 221-227.

[32] K. Makarananda *et al.*, Monitoring of Aflatoxin Exposure by Biomarkers, *J. Toxicol. Sci.* 123 (1998) 155-9.

[33] C. Chen *et al.*, Chronic Hepatitis B Carriers with Null Genotypes of Glutathione *S*-transferase M1 and T1 Polymorphisms Who are Exposed to Aflatoxin are at Increased Risk of Hepatocellular Carcinoma, *Am. J. Hum. Genet.* 59 (1996) 128-134.

[34] E. Loechler, The Role of Adduct Site Specific Mutagenesis in Understanding How Carcinogen-DNA Adducts Can Cause Mutations: Perspectives, Prospects and Problems, *Carninogenesis* 17 (5) (1996) 895-902.

[35] K. Hemminki, DNA Adducts and Mutations in Occupational and Environmental Biomonitoring, *Environ. Health Perspect.* 105(Suppl 4) (1997) 823-827.

[36] O. Froment *et al.*, Mutagenesis *ras* Proto-oncogenes in Rat Liver and Cancers Induced by Vinyl Chloride, *Cancer Res.* 54(20) (1994) 5340-5.

[37] R. Misra *et al.*, A Comparison of Two Ultrasentitive Methods for Measuring 1, N6-etheno 2′ deoxyadenosine and 3, N4-etheno 2′ Deoxycytidine in Cellular DNA, *Carcinogenesis* 15 (1994) 1647-1652.

[38] C. Zhao *et al.*, Ultraviolet Photoproduct Levels in Melanocytic Nevi and Surrounding Epidermis in Human Skin *In Situ*, *J. Invest. Dermatol.* 118(1) (2002) 180-4.

[39] H. Ahsan *et al.*, Variability in Aflatoxin-Albumin Adduct Levels and Effects of Hepatitis B and C Virus Infection and Glutathione *S*-transferase *M1* and *T1* Genotype, *Environ. Health Perspect.* 109 (2001) 833-837.

[40] M. King *et al.*, Immunoaffinity Concentration of Human DNA Adducts Using an Anti-benzo[a]pyrene-diol-epoxide-DNA Antibody. Analysis by ^{32}P-postlabelling or ELISA, *Mut. Res.* 292 (1993) 113-122.

[41] E. Hoevath *et al.*, ^{32}P-Post-labeling Detection of DNA Adducts in Mononuclear Cells of Workers Occupationally Exposed to Styrene, *Carcinogenesis* 15(7)(1994) 1309-1315.

[42] R. Zielhuis, Theoretical and Practical Consideration in Biological Monitoring, In: A. Aitio et al., (eds), Biological Monitoring and Surveillance of Workers Exposed to Chemicals, New York: Hemisphere Publishing, 1984, pp. 7-18.

[43] H. Hernberg. Lead. In: A. Aitio et al., (eds), Biological Monitoring and Surveillance of Workers Exposed to Chemicals, New York: Hemisphere Publishing, 1984, pp. 27-35.

[44] S. Pichini *et al.*, Cord Serum Cotinine as a Biomarker of Fetal Exposure to Cigarette Smoke at the End of Pregnancy, *Environ. Health Perspect.* 108 (2000) 1079-1083.

[45] D. Anderson *et al.*, The Parallelogram Approach in Studies of Genotoxic Effects, *Mut. Res.* 313 (1994) 101-115.

[46] J. Ward and R. Henderson, Identification of Needs in Biomarker Research. *Environ. Health Perspect.* 104 (1996) 895-900.

[47] F. Lowry, Cytokine Mediators of Immunity and Inflammation, *Arch. Surg.* 8 (1993) 1235-1241.

[48] H. Lander, An Essential Role for Free Radicals and Derived Species in Signal Transduction, FASEB 11 (1997) 118-124.

[49] P. Cerutti, Prooxidant States and Tumor Promotion, Science 227(1985) 375-381.

[50] H. Wiesman and B. Halliwell, Damage to DNA by Reactive Oxygen Species and Nitrogen Species: Role of Inflammatory Disease and Progression to Cancer, *Biochem. J.* 313 (1996) 17-29.

[51] B. Freeman and J. Crapo, Biology of Disease: Free Radicals and Tissue Injury, *Lab. Invest.* 47 (1982) 412 – 426.

[52] J. Fanton and P. Ward, Role of Oxygen Derived Free Radicals and Metabolites in Leukocyte Dependent Inflammatory Reactions, Am. J. Pathol. 107(1982) 398-418.

[53] United Nations Scientific Committee on the Effects of Atomic Radiation. 1994 Report to the General Assembly, With Annexes. Sources and Effects of Ionizing Radiation. New York:United Nations, 1994.

[54] S. Wolff et al.,. Human Lymphocytes Exposed to Low Doses of Ionizing Radiations Become Refractory to High Doses of Radiation as well as to Chemical Mutagens that Induce Double-strand Breaks in DNA, *Int. J. Radiat. Biol.* 53 (1998) 39-47.

[55] M. Cosma, Ordered Recruitment: Gene-specific Mechanism of Transcription Activation, Mol. Cell 10(2) (2002) 227-36.

[56] G. Chu and E. Chang, Xeroderma Pigmentosum Group E Cells Lack a Nuclear Factor that Binds to Damaged DNA, *Science* 242 (1988) 564- 567.

[57] A. Whitmarsh and R. Davis, Regulation of Transcription Factor Function by Phosphorylation, *Cell Mol. Life Sci.* 57 (2000) 1172-1183.

[58] R. Sen and D. Baltimore, Multiple Nuclear Factors Interact with the Immunoglobulin Enhancer Sequences, *Cell* 46 (1986) 705-716.

[59] L. Escoubet-Lozach *et al.*, The Role of Transcription Factors In Allergic Inflammation, Allergy Clin. Immunol. 110(4) (2002) 553-64.

[60] K. Yasumoto *et al.*, Tumor Necrosis Factor Alpha and Interferon Gamma Synergistically Induce Interleukin-8 Production in a Human Gastric Cancer Cell Line Through Acting Concurrently on AP-1 and NF-kB -like Binding Sites of the Interleukin-8 Gene, *J. Biol. Chem.* 267 (1992) 22506-22511.

[61] A. Roebuck, Oxidant Stress Regulation of IL-8 and ICAM-1 Gene Expression: Differential Activation and Binding of the Transcription Factors AP-1 and NF-kappaB, *Int. J. Mol. Med.* 4 (1999) 223-230.

[62] P. Baeuerle and T. Henkel, Function and Activation of NF-κB in the Immune System, *Annu. Rev. Immunol.* 12 (1994) 141-179.

[63] A. Baldwin, The NF-κB and IκB Proteins: New Discoveries and Insights, *Annu. Rev. Immunol.* 14 (1996) 649-681.

[64] P. Baeuerle and D. Baltimore, IκB: A Specific Inhibitor of the NF-κB Transcription Factor, *Science* 242 (1988) 540-546.

[65] U. Siebenlist *et al.*, Structure, Regulation and Function of NF-κB, *Annu. Rev. Cell Biol.* 10 (1994) 405-455.

[66] S. Ghosh *et al.*, NF-κB and Rel Proteins: Evolutionarily Conserved Mediators of Immune Responses, *Annu. Rev. Immunol.* 16 (1998) 225-260.

[67] P. Cogswell *et al.*, Involvement of Egr-1/RelA Synergy in Distinguishing T Cell Activation from Tumor Necrosis Factor-ά-induced NF-kB1 Transcription, *J. Exp. Med.* 195 (1997) 491-498.

[68] R. Costello *et al.*, The Role of NF-KB1 (p50/p105) Gene Expression in Activation of Human Blood T-lymphocytes via CD2 and CD28 Adhesion Molecules, *Cell Growth Differ.* 4 (1993) 947-954.

[69] C. Saliou *et al.*, Antioxidants Modulate Acute Solar Ultraviolet Radiation-Induced NF-kappa-B Activation in a Human Keratinocyte Cell Line, *Free Radiat. Biol. Med.* 26(1-2) (1999) 174-83.

[70] E. Haddad *et al.*, Ozone Induction of Cytokine-induced Neutrophil Chemoattractant (CINC) and Nuclear Factor-kappa B in Rat Lung: Inhibition by Corticosteroids, *FEBS Lett.* 379 (1996) 265-268.

[71] W. Joshua *et al.*, Molecular Basis for Effects of Carcinogenic Heavy Metals on Inducible Gene Expression, *Environ. Health Perspect.* 106(Suppl 4) (1998) 1005-1015.
[72] Z. Dong, The Molecular Mechanisms of Arsenic-Induced Cell Transformation and Apoptosis, *Environ. Health Perspect.* 110(suppl 5) (2002) 757-759.
[73] S. Faux and J. Howden., Possible Role of Lipid Peroxidation in the Induction of NF- κB and AP-1 in RFL-6 Cells by Crocidolite Asbestos: Evidence following Protection by Vitamin E, *Environ. Health Perspect.* 105(Suppl 5) (1997) 1127-1130.
[74] A. Iarilin, Radiation and Immunity. Interference of Ionizing Radiation with Key Immune Processes, *Radiat. Biol. Radioecol.* 39(1) (1999) 181-189.
[75] J. Li *et al.*, Inhibitors of Both Nuclear Factor- κB and Activator Protein-1 Activation Block the Neoplastic Transformation Response, *Cancer Res.* 57 (1997) 3569-3576.
[76] H. Wu and G. Lozano, NF-κB Activation of p53: A Potential Mechanism for Suppressing Cell Growth in Response to Stress, *J. Biol. Chem.* 269 (1994) 20067-20074.
[77] P. Angel and M. Karin, The Role of Jun, Fos and the AP-1 Complex in Cell Proliferation and Transformation, *Biochim. Biophys. Acta* 1072 (1991) 129-157.
[78] L. Ransone and I. Verma, Nuclear Proto-oncogenes Fos and Jun, *Annu. Rev. Cell Biol.* 6 (1990) 539-557.
[79] M. Mietus-Snyder *et al.*, Transcriptional Activation of Scavenger Receptor Expression in Human Smooth Muscle Cells Requires AP-1/c-Jun and C/EBP: Both AP-1 Binding and JNK Activation Are Induced by Phorbol Esters and Oxidative Stress, *Arterioscler. Thromb. Vasc. Biol.* 18(9) (1998) 1440-1449.
[80] S. Pandey and E. Wang, Cells En Route to Apoptosis are Characterized by the Upregulation of c *-fos* , c *-myc* , c *-jun* , cdc2, and RB Phosphorylation, Resembling Events of Early Cell-cycle Traverse, *J Cell Biochem.* 58 (1995) 135-150.
[81] K. Suzukawa *et al.*, AP-1, NFkB, and ERK Activation: Thresholds for Promotion of Neoplastic Transformation in the Mouse Epidermal JB6 Model, *Environ. Health Perspect.* 110(9) (2000) 865-870.
[82] Z. Dong *et al.*, Blocking of Tumor Promoter-induced AP-1 Activity Inhibits Induced Transformation in JB6 Mouse Epidermal Cells, *Proc. Natl. Acad. Sci. USA* 91 (1994) 609-613.
[83] M. Young *et al.*, Transgenic Mice Demonstrate AP-1 (Activator Protein-1) Transactivation Is Required for Tumor Promotion, *Proc. Natl. Acad. Sci. USA* 96 (1999) 9827-9832.
[84] C. Timblin *et al.*, Transcriptional Activation of the Proto-oncogene c- *jun* , by Asbestos and H_2O_2 Is Directly Related to Increased Proliferation and Trans-formation of Tracheal Epithelial Cells, *Cancer Res.* 55 (1995) 2723-2726.
[85] D. Frigo *et al.*, DDT and Its Metabolites Alter Gene Expression in Human Uterine Cell Lines Through Estrogen Receptor-Independent Mechanisms, *Environ. Health Perspect.* 110 (2002) 1239-1245.
[86] W. Chen *et al.*, UVB Irradiation-induced Activator Protein-1 Activation Correlates with Increased c-*fos* Gene Expression in a Human Keratinocyte Cell Line, *J. Biol. Chem.* 273(48) (1998) 32176-32181.
[87] I. Ferrer *et al.*, Radiation-induced Apoptosis in Developing Rats and Kainic Acid-induced Excitotoxicity in Adult Rats Are Associated with Distinctive Morphological and Biochemical C-Jun/AP-1 Expression, *Neuroscience* 80(2) (1997) 449-58.
[88] N. Chen *et al.*, Inhibition of Arsenite-induced Apoptosis and AP-1 Activity by Epigallocatechin-3-gallate and Theaflavins, *J. Environ. Pathol. Toxicol. Oncol.* 19(3) (2000) 287-295.
[89] C. Huang *et al.*, Transactivation of AP-1 in AP-1-luciferase Reporter Transgenic Mice by Arsenite and Arsenate, *Anticancer Res.* 21 (2001) 261-268.
[90] E. Brugnera *et al.*, Chromosomal Mapping and Characterization of the Human Metal-regulatory Transcription Factor MTF-1, *Nucleic Acids Res.* 22 (1994) 3167-3173.
[91] R. Heuchel *et al.*, The Transcription factor MTF-1 is Essential for Basal and Heavy Metal-induced Metallothionein Gene Expression, *EMBO J.* 13 (1994) 2870-2875.
[92] D. Hamer, Metallothionein, *Annu. Rev. Biochem.* 55 (1986) 913-951.
[93] J. Kagi and A. Schaffer, Biochemistry of Metallothionein, *Biochemistry* 27 (1988) 8509-8515.
[94] P. Thornalley and M.Vasak, Possible Role for Metallothionein in Protection Against Radiation-induced Oxidative Stress. Kinetics and Mechanism of its Reaction with Superoxide and Hydroxyl Radicals, *Biochim. Biophys. Acta* 827 (1985) 36-44.
[95] N. Saydam *et al.*, Regulation of Metallothionein Transcription by the Metal-Responsive Transcription Factor MTF-1. Idetification of Signal Transduction Cascades that Control Metal-inducible Transcription, *J. Biol. Chem.* 277(23) (2002) 20438-20445.
[96] O. LaRochelle *et al.*, Phosphorylation Is Involved in the Activation of Metal-Regulatory Transcription Factor 1 in Response to Metal Ions, *J. Biol. Chem.* 276(45) (2001) 41879-41888.
[97] F. Radtke *et al.*, Functional Domains of the Heavy Metal-responsive Transcription Regulator MTF-1. *Nucleic Acids Res.* 23 (1995) 2277-2286.
[98] I. Smirnova *et al.*, Zinc and Cadmium Can Promote Rapid Nuclear Translocation of Metal Response Element-binding Transcription Factor-1, *J. Biol. Chem.* 275 (13) (2000) 9377-9384.

[99] K. Timothy *et al.*, Activation of Gene Expression by Metal-Responsive Signal Transduction Pathways, *Environ. Health Perspect.* 110(suppl 5) (2002) 813-817.
[100] K. Ghoshal *et al.*, Virus Infection Induces Metallothionein Gene Expression in the Mouse Liver and Lung by Overlapping but Distinct Molecular Mechanisms, *Mol. Cell Biol.* 21(24) (2001) 830-8317.
[101] T. Dalton *et al.*, Oxidative Stress Activates Metal-responsive Transcription Factor-1 Binding Activity; Occupancy *In Vivo* of Metal Response Elements in the Metallothionein-1 Gene Promotor, *J. Biol. Chem.* 271(42) (1996) 26233-26241.
[102] G. Murphy *et al.*, Activation of Metallothionein Gene Expression by Hypoxia Involves Metal Response Elements and Metal Transcription Factor-1, *Cancer Res.* 59(6) (1999) 1315-1322.
[103] I. Fridovich, Superoxide Radical and Superoxide Dismutases, *Ann. Rev. Biochem.* 64 (1995) 97-112.
[104] L. Fialkow *et al.*, Regulation of Tyrosine Phosphorylation in Neutrophils by the NADPH Oxidase.Role of Reactive Oxygen Intermediates, *J. Biol. Chem.* 268 (1993) 17131-17137.
[105] S. Weiss, Oxygen, Ischemia and Inflammation, *Acta Physiol. Scand. Suppl.* 548 1986) 9-37.
[106] S. James, Role of Nitric Oxide in Parasitic Infection, *Microbiological Rev.* 59(4) (1995) 533-547.
[107] T. Dalton *et al.*, Regulation of Gene Expression by Reactive Oxygen, *Annu. Rev. Pharmacol. Toxicol.* 39(1) (1999) 67-101.
[108] T. Storch and G. Talley, Oxygen Concentration Regulates the Proliferative Response of Fibroblasts to Serum and Growth Factors, *Exp. Cell Res.* 175 (1988) 317-325.
[109] T. Buttke and P. Sandstrom, Oxidative Stress as a Mediator of Apoptosis. *Immunol. Today* 15(1994) 7-10.
[110] C. Sen and L. Packer, Antioxidant and Redox Regulation of Gene Transcription, *FASEB J.* 10 (1996) 709-720.
[111] C. Vallyathan and X. Shi, The Role of Oxygen Free Radicals in Occupational and Environmental Lung Diseases, *Environ. Health Perspect.* 105(Suppl 1) (1997) 165-177
[112] P. Henson and R. Johnston, Tissue Injury in Inflammation, Oxidants, Proteinases and Cationic Proteins, *J. Clin. Invest.* 79 (1987) 669-674.
[113] B. Ames, Dietary Carcinigens and Anticarciniogens, *Science* 121 (1983) 1250-1264.
[114] P. Early and D. Sodee, In: Principles and Practice of Nuclear Medicine, C.V. Mosby Company, St, Louis, Toronto, 1985, pp. 4-12.
[115] D. Valerie *et al.*, Oxygen Radicals and Asbestos, Carcinogenesis, *Environ. Health Perspect.* 102(suppl 10) (1994) 131-136.
[116] E. Fehrenbach and H. Northoff, Free Radicals, Exercise, Apoptosis, and Heat Shock Proteins, *Exerc. Immunol. Rev.* 7 (2001) 66-89.
[117] C. Cross *et al.*, Oxygen Radical and Human Disease, *Ann. Intern. Med.* 107 (1987) 526-545.
[118] P. Southorn and G. Powis, Free Radicals in Medicine. II. Involvement in Human Disease, *Mayo Clin. Proc.* 63 (1988) 390-408.
[119] B. Halliwell, Free Radicals, Antioxidants, and Human Disease: Curiosity, Cause or Consequence? *Lancet* 344 (1994) 721-724.
[120] C. Nathan and Q. Xie, Nitric Oxide Synthetases: Roles, Tolls and Controls, *Cell* 78 (1994) 915-918.
[121] C. Nathan, Nitric Oxide as a Secretory Product of Mammalian Cells, *FASEB J.* 6 (1992) 3051-3064.
[122] S. Gross and M. Wolin, Nitric oxide: Pathophysiological Mechanisms, *Ann. Rev. Physiol.* 57 (1995) 737-769.
[123] G. Clough, Role of Nitric Oxide in the Regulation of Microvascular Perfusion in Human Skin *In Vivo, J. Physiol.* 516 (1999) 549-557.
[124] Q. Yang and G. Hatton, Histamine H(1)-receptor Modulation of Inter-neuronal Coupling Among Vasopressinergic Neurons Depends on Nitric Oxide Synthase Activation, *Brain Res.* 955(1-2) (2002) 115-22.
[125] J. MacMicking *et al.*, Nitric Oxide and Macrophage Function, *Ann. Rev. Immunol.* 15 (1997) 323-350.
[126] B. Stallmeyer *et al.*, The Function of Nitric Oxide in Wound Repair: Inhibition of Inducible Nitric Oxide-synthase Severely Impairs Wound Reepithelialization, *J. Invest. Dermatol.* 113 (1999) 1090-1098.
[127] J. Reichner *et al.*, Molecular and Metabolic Evidence for the Restricted Expression of Inducible Nitric Oxide Synthase in Healing Wounds, *Am. J. Pathol.* 154 (1999) 1097-1104.
[128] S. Lee *et al.*, Protective Role of Nitric Oxide-mediated Inflammatory Response Against Lipid Peroxidation In Ultraviolet B-irradiated Skin, *Br. J. Dermatol.* 142(4) (2000) 653-659.
[129] W. MacNaughton *et al.*, Expression, Activity and Cellular Localization of Inducible Nitric Oxide Synthase in Rat Ileum and Colon Post-irradiation, *Int. J. Radiat. Biol.* 74(2) (1998) 255-64.
[130] M. El Batanouny and S. Korraa, Effects OF Low Intensity Laser on the Activity and Expression of Nitric Oxide Synthase In Human Polymorphonuclear Leucocytes *In Vitro, Arab J. Lab. Med.* 28(3) (2002) 289-297.

[131] A. Marrogi *et al.*, Human Mesothelioma Samples Overexpress Both Cyclooxygenase-2 (COX-2) and Inducible Nitric Oxide Synthase (NOS2): *In Vitro* Antiproliferative Effects of a COX-2 Inhibitor, *Cancer Res.* 60(14) (2000) 3696-3700.
[132] J. Smith Sonneborn, Stress Protein Response in Hormesis. In: Biological Effects of Low Level Exposures to Chemicals and Radiation, E. Calabrese (ed), Chelsea, MI:Lewis 1992, pp. 41-52.
[133] P. Brandt-Rauf et al., Serum Oncoproteins and Growth Factors in Asbestosis and Silicosis Patients, *Int. J. Cancer,* 50 (1992) 881-885.
[134] M. Nguyen *et al.*, Elevated Levels of an Angiogenic peptide, Basic Fibroblast Growth Factor, in the Urine of Patients with a Wide Spectrum of Cancers, *J. Natl. Cancer Inst.* 86 (1994) 356-362.
[135] P. Brandt-Rauf *et al.*, The Molecular Epidemiology of Growth Signal Transduction Proteins, *J. Occup. Environ. Med.* 37 (1995) 77-83.
[136] T. Enoch and C. Norbury, Cellular Responses to DNA Damage: Cell-cycle Checkpoints, Apoptosis and the Roles of p53 and ATM. *Trends Biochem. Soc.* 20 (1995) 426-430.
[137] A. Braun *et al.*, Detection of Oncogene and Tumor Suppressor Gene Products in Serum of Former Uranium Miners for Secondary Prevention of Radon-Induced Lung Cancer. *Proc. Eur. Con. Biomarkers Environ. Toxicol.* (1995) 2:A6.
[138] P. Brandt-Rauf, Biomarkers of Gene Expression: Growth Factors and Oncoproteins, *Environ. Health Perspect.* 105(Suppl 4) (1997) 807-816.
[139] D. Steel, The Role of Growth Factors in Wound Healing, *Surgical Clinics of North America,* 77(3) (1997) 575-585.
[140] M. Pawlikowski *et al.*, Elevated Blood Serum levels of Epidermal Growth Factor in Some Patients with Gastric Cancer, *Endokrynol. Pol.* 40 (1989) 149-153. (Abs.)
[141] J. Bhatavdekar *et al.*, Circulating Markers and Growth Factors as Prognosticators in Men with Advanced Tongue Cancer, *Tumor Biol.* 14 (1993) 55-58.
[142] N. Shah *et al.*, Circulating EGF and IGF-I in Patients with Epithelial Ovarian Carcinoma, *Neoplasma* 41 (1994) 241-243.
[143] S. Korraa *et al.*, Biomonitoring of Pesticides Formulators: Frequency of Micronuclei, Protein Carbonyls, Epidermal Growth Factor, and Basic Fibroblast Growth Factor, *Medical J. of Cairo Univ.* 54 (2003) 112-119.
[144] B. Lauber *et al.*, An Autocrine Mitogenic Activity Produced by a Pleural Human Mesothelioma Cell Line, *Int. J. Cancer* 50 (1992) 943-950.
[145] J. Peyrat *et al.*, Insulin-like Growth Factor 1 (IGF1) Plasma Concentration in Breast Cancer Patients, *Proc. Am. Assoc. Cancer Res.* 31 (1990) A295.
[146] S. Rojdmark *et al.*, Insulin-like Growth Factor (IGF)-1 and IGF-binding Protein-1 Concentrations in Serum of Normal Subjects after Alcohol Ingestion: Evidence for Decreased IGF-1 Bioavailability, *Clin. Endocrinol.* 14 (2000) 313-8.
[147] J. Bonner and A. Brody, Asbestos Induced Alveolar Injury; Evidence for Macrophage-derived PDGF as a Mediator of the Fibrogenic Response, *Chest* 99 (1991) 545-555.
[148] D. Germolec, Arsenic Enhancement of Skin Neoplasia by Chronic Stimulation of Growth Factors, *Am. J. Pathol.* 153 (1998) 1175-1185.
[149] D. Aebersold *et al.*, Expression of Transforming Growth Factor-alpha, Epidermal Growth Factor Receptor and Platelet-derived Growth Factors A and B in Oropharyngeal Cancers Treated by Curative Radiation Therapy, *Radiother. Oncol.* 63 (3) (2002) 275-83.
[150] O. Hioki *et al.*, Clinical Significance of Serum Hepatocyte Growth Factor Levels in Liver Diseases, *J. Med.* 24 (1993) 5-46.
[151] J. Schneider *al.*, p53 Protein, EGF receptor, and Anti-p53 Antibodies in Serum From Patients with Occupationally Derived Lung Cancer, *Br. J. Cancer* 80(12) (1999) 1987-94.
[152] R. Partanen *et al.*, The Detection of Increased Amounts of the Extracellular Domain of the Epidermal Growth Factor Receptor in Serum During Carcinogenesis in Asbestosis Patients, *J. Occup. Med.* 36 (1994) 1324-1328.
[153] R. Partanen *et al.*, Serum Levels of Growth Factor Receptors, EGFR and Neu in Asbestosis Patients: a Follow-up Study, *Int. J. Oncol.* 4 (1994) 1025-1028.
[154] L. Witters *et al.*, Epidermal Growth factor Receptor Ectodomain in the Urine of Patients with Squamous Cell Carcinoma. *Clin. Cancer Res.* 1 (1995) 551-557.
[155] A. Tritcher *et al.*, Dose-Response Effects of Dioxin- Species Comparisons and Implications for Risk Assessment, In: A. Shecter (ed.), Dioxins and Health. ISBN 0-306-44785-1. Plenum Press, New York, 1994, pp. 227-247.
[156] S. Smith *et al.*, Quantitation of the Extracellular Domainof Epidermal Growth Factor Receptor in the Plasma of Dioxin-Exposed Individuals, Am. *J. Ind. Med.* 34 (1998) 1-5.
[157] C. Sherr, Mammalian G1 Cyclins, *Cell* 73 (1993) 1059-1065.
[158] R. Weinberg, The Retinoblastoma Protein and Cell Cycle Control, *Cell* 81 (1995) 323-330.

[159] H. Matsushime *et al.*, Colony-stimulating Factor 1 Regulates Novel Cyclins During the G1 Phase of the Cell Cycle, *Cell* 65 (1991) 701-713.

[160] H. Matsushime *et al.*, Identification and Properties of an Atypical Catalytic Subunit (p34PSK-J3/Cdk4) for Mammalian D-type G1 Cyclins, *Cell* 71 (1992) 323-334.

[161] Y. Xiong *et al.*, D type Cyclins Associate with Multiple Protein Kinases and the DNA Replication and Repair Factor PCNA, *Cell* 71 (1992) 505-51.

[162] D. Quelle *et al.*, Overexpression of Mouse D-type Cyclins Accelerates G1 Phase in Rodent Fibroblasts, *Genes & Devel.* 7 (1993) 1559-1571.

[163] L. Hartwell and M. Kastan, Cell Cycle Control and Cancer, *Science* 266 (1994) 1821-1828.

[164] J. Brugarolas *et al.*, Radiation-induced Cell Cycle Arrest Compromised by p21 Defficiency, *Nature* 377 (1995) 552-55.

[165] V. Levresse *et al.*, Analysis of Cell Cycle Disruptions in Cultures of Rat Pleural Mesothelial Cells Exposed to Asbestos Fibers, *Am. J. Respir. Cell Mol. Biol.* 17(6) (1997) 660-671.

[166] I. Adamson, Early Mesothelial Cell Proliferation after Asbestos Exposure: *In Vivo and in Vitro* Studies, *Environ. Health Perspect.* 105(Suppl 5) (1997) 1205-1208.

[167] K. Dhanwada *et al.*, Differential Effects of UV-B and UV-C Components of Solar Radiation on MAP Kinase Signal Transduction Pathways in Epidermal Keratinocytes, *Oncogene* 11(10) (1995) 1947-53.

[168] C. Götz and M. Montenahr, P53: DNA Damage, DNA Repair and Apoptosis. *Rev. Physiol. Biochem. Pharmacol.* 127 (1993) 65-95.

[169] R. Shackelford *et al.*, Cell Cycle Control, Checkpoint Mechanisms, and Genotoxic Stress, *Environ. Health Perspect.* 107(Suppl 1) (1999) 5-25.

[170] T. Krontiris, Oncogenes, *N. Engl. J. Med.* 333 (1995) 303.

[171] M. Barbacid, Ras Genes, *Ann. Rev. of Biochem.* 56 (1987) 779-827.

[172] M.J. Marion *et al.*, Activationn of Ki-ras Gene by Point Mutations in Human Liver Angiosarcoma Associated with Vinyl Chloride Exposure, *Molecular Carcinogenesis* 4 (1991) 450-454.

[173] D. Anderson *et al.*, Levels of Ras Oncoproteins in Human Plasma 1,3-Butadiene-exposed Workers and Controls, *Mut. Res.* 349 (1996) 115-120.

[174] M. Corominas *et al.*, Ras Activation in Human Tumors and in Animal Model Systems, *Environ. Health Perspect.* 93 (1991) 19-25.

[175] G. **Fritz** *et al.*, The Ras-related Small GTP-binding Protein RhoB Is Immediate-early Inducible by DNA Damaging treatments, *Biol. Chem.* 270(42) (1995) 25172-7.

[176] M. Iordanov *et al.*, The UV (Ribotoxic) Stress Response of Human Keratinocytes Involves the Unexpected Uncoupling of the Ras-extracellular Signal-regulated Kinase Signaling Cascade from the Activated Epidermal Growth Factor Receptor, *Mol. Cell Biol.* (15) (2002) 5380-94.

[177] M. Grunstein *et al.*, Rhinovirus Elicits Proasthmatic Changes in Airway Responsiveness Independently of Viral Infection, *Allergy Clin. Immunol.* 108(6) (2001) 997-1004.

[178] S. Kim *et al.*, Retinoblastoma Gene Product Activates Expression of the Human TGF-ß2 Gene through Transcription Factor ATF-2, *Nature* 358 (1992) 331-334.

[179] W. Border and A. Noble, Transforming Growth Factor ß in Tissue Fibrosis, *N. Engl. J. Med.* 331 (1994) 1286-1292.

[180] J. Massagué, The Transforming Growth Factor-ß Family, *Annu. Rev. Cell Biol.* 6 (1990) 597-641.

[181] S. Wahl, Transforming Growth Factor ß (TGF-ß) in Inflammation: a Cause and a Cure, *J. Clin. Immunol.* 12 (1992) 61-74.

[182] R. Derynck *et al.*, TGF-beta Signaling in Tumor Suppression and Cancer Progression, *Nat. Genet.* 29(3) (2001) 351.

[183] A. Roberts *et al.*, TGF-ß: Regulation of Extracellular Matrix, *Kidney Int.* 41(1992) 557-559.

[184] T. Parekh *et al.*,.Transforming Growth Factor Beta Signaling is Disabled Early in Human Endometrial Carcinogenesis Concomitant with Loss of Growth Inhibition, *Cancer Res.* 62(10) (2002) 2778-90.

[185] S. Kim *et al.*, Transforming Growth Factor Expression in Normal and Pathological Conditions, *Horm. Res.* 42 (1994) 5-8.

[186] J. Jagirdar *et al.*, Immunohistochemical Localization of Transforming Growth Factor Beta Isoforms in Asbestos-Related Diseases, *Environ. Health Perspect.* 105(Suppl 5) (1997) 1197-1203.

[187] R. Neta, Modulation with Cytokines of Radiation Injury: Suggested Mechanisms of Action, *Environ. Health Perspect.* 105(Suppl 6) (1997) 1463-1465.

[188] K. Danno *et al.*, Near-infrared Irradiation Stimulates Cutaneous Wound Repair: Laboratory Experiments on Possible Mechanisms, *Photodermatol. Photoimmunol. Photomed.* 17(6) (2001) 261-5.

[189] T. Quan *et al.*, Ultraviolet Irradiation Alters Transforming Growth Factor Beta/smad Pathway in Human Skin *In Vivo*, *J. Invest. Dermatol.* 119(2) (2002) 499-506.

[190] A. Haugen *et al.*, Nickel-induced Alterations in Human Renal Epithelial Cells, *Environ. Health Perspect.* 102(Suppl 3) (1994) 117-118.

[191] N. Azam *et al.*, Interaction of CR6 (GADD45Gamma) with Proliferating Cell Nuclear Antigen Impedes Negative Growth Control, *Biol. Chem.* 276(4) (2001) 2766-74.

[192] V. Masciullo *et al.*, The Rb Family of Cell Cycle Regulatory Factors: Clinical Implications, *Int. J. Oncol.* 17(5) (2000) 897-902.

[193] Y. Chen *et al.*, Tumor Suppression by p21WAF1. *Cancer Res.* 1995 55(20) (1995) 536-9.

[194] B. Vogelstein and K. Kinzler, p53 function and dysfunction, *Cell* 70 (1992) 523-526.

[195] C. Harris and M. Hollstein, Clinical Implications of the p53 Tumor Suppressor Gene, *N. Engl. J. Med.* 329 (1993) 1318-1327.

[196] M. Kastan *et al.*, p53, Cell Cycle Control and Apoptosis: Implications for Cancer, *Cancer Metastasis Rev.* 14 (1995) 3-15.

[197] J. Lee and A. Bernstein, Apoptosis, Cancer and the p53 Tumour Suppressor Gene, *Cancer Metastasis Rev.* 14 (1995) 149-161.

[198] M. Hollstein *et al.*, p53 Mutations in Human Cancer, *Science* 253 (1991) 49-53 (1991).

[199] M. Greenblatt *et al.*, Mutations in the p53 Tumor Suppressor Gene: Clues to Cancer Etiology and Molecular Pathogenesis, *Cancer Res.* 54 (1994) 4855-4878.

[200] T. Lasky and E. Silbergeld, p53 Mutations Associated with Breast, Colorectal, Liver, Lung, and Ovarian Cancers, *Environ. Health Perspect.* 104 (1996) 1324-1331.

[201] S. Hanaoka *et al.*, Elevated Serum levels of Pantropic p53 Proteins in Chromium Workers, *Scand. J. of Work, Environ. Health* 23 (1997) 37-40.

[202] G. Trivers *et al.*, Anti-p53 Antibodies in Sera of Workers Occupationally Exposed to Vinyl Chloride, *J. Natl. Cancer Inst.* 87 (1995) 1400-1407.

[203] F. Aguilar *et al.*, Afatoxin B_1 Induces the Transversion of G→T In Codon 249 of the p53 Tumor Suppressor Gene in Human Hepatocytes, *Proc. Natl. Acad. Sci. USA* 90 (1993) 8586-8590.

[204] J. Trosko *et al.*, Modulators of Gap Junction Function: the Scientific Basis of Epigenetic Toxicology, *In Vitro Toxicol.* 3 (1990) 9-26.

[205] J. Trosko *et al.*, *In Vitro* Analysis of Modulators of Intercellular Communication: Implications for Biologically Based Risk Assessment Models for Chemical Exposure, *Toxicol In Vitro* 4 (1990) 635-643.

[206] J. Trosko *et al.*, The Role of Modulated Gap Junctional Intercellular Communication in Epigenetic Toxicology, *Risk Anal.* 14(3) (1994) 303-312.

[207] J. Trosko *et al.*, Endogenous and Exogenous Modulation of Gap Junctional Intercellular Communication: Toxicological and Pharmacological Implications, *Life Sci.* 53 (1993) 1-19.

[208] H. Yamasaki and C. Naus, Role of Connexin Genes in Growth Control, *Carcinogenesis* 17 (1996) 1199-1213.

[209] V. Krutovskikh *et al.*, Differential Effect of Subcellular Localization of Communication Impairing Gap Junction Protein Connexin43 on Tumor Cell Growth *In Vivo*, *Oncogene* 19(4) (2000) 505-13.

[210] Y. Omori *et al.*, Role of Connexin (Gap Junction) Genes in Cell Growth Control: Approach with Site-directed Mutagenesis and Dominant-negative Effects, *Toxicol. Lett.* 96-97 (1998) 105-10.

[211] C. Mothersill and C. Seymour, Cell-cell Contact During Gamma Irradiation is Not Required to Induce a Bystander Effect in Normal Human Keratinocytes: Evidence for Release During Irradiation of a Signal Controlling Survival Into the Medium. Radiat Res 1998 Mar;149(3):256-62.

[212] M. Liliane *et al.*, Bay or Baylike Regions of Polycyclic Aromatic Hydrocarbons Were Potent Inhibitors of Gap Junctional Intercellular Communication, *Environ. Health Perspect.* 106 (1998) 17-22.

[213] K. Kang *et al.*, Inhibition of Gap Junctional Intercellular Communication in Normal Human Breast Epithelial Cells After Treatment with Pesticides, PCBs, and PBBs, Alone or in Mixtures, *Environ. Health Perspect.* 104(2) (1996) 192-200.

[214] H. Banrud *et al.*, Effects of Ultraviolet Radiation on Intercellular Communication in V79 Chinese Hamster Fibroblasts, *Carcinogenesis* 15(2) (1994) 233-9.

[215] D. Yamaguchi *et al.*, Inhibition of Gap Junction Intercellular Communication by Extremely Low-frequency Electromagnetic Fields in Osteoblast-like Models is Dependent on Cell Differentiation, *Cell Physiol.* 190(2) (2000) 180-8.

[216] R. Nakaoka *et al.*, The Inhibitory Mechanism of Gap Junctional Intercellular Communication Induced by Polyethylene and the Restorative Effects by Surface Modification with Various Proteins. *J Biomed. Mater. Res.* 57(4) (2001) 567-74.

[217] S. McKarns and D. Doolittle, Quantitative Approach to Assessing Intercellular Communication: Studies on Cigarette Smoke Condensates, *Toxico. Appl. Pharmacol.* 111(1) (1991) 58-68.

[218] T. Ikushima, Radio-adaptive Response: Involvement of Induction of Specific Gene Expression by Low Doses of Ionizing Radiation. In: Low Dose Irradiation and Biological Defense Mechanisms (T. Sugahara *et al.*, eds). Amsterdam:Elsevier, 1992, pp. 255-258.

[219] N. Maulik *et al.*, Myocardial Adaptation to Ischemia by Oxidative Stress Induced by Endotoxin, *Am. J. Physiol.* 269 (1995) 907-916.

[220] R. VanBogelen *et al.*, Differential Induction of Heat Shock, SOS and Oxidation Stress Regulons and Accumulation of Nucleotides in Escherichia Coli, *Bacteriol.* 169 (1987) 26-32.

[221] G. Balla *et al.*, Ferritin a Cytoprotective Antioxidant Stratagems of Endothelium, *J. Biol. Chem.* 267 (1992) 1-6.
[222] S. Cregan *et al.*, Apoptosis and the Adaptive Response in Human Lymphocytes, *Int. J. Radiat. Biol.* 75(9) (1999) 1087-94.
[223] P. Amstad and P. Cerutti, Genetic Modulation of the Cellular Antioxidant Defense Capacity, *Environ. Health Perspect.* 88 (1990) 77-82.
[224] R. Persinger *et al.*, 'Differences in Constitutive and Inducible Expression of Antioxidant Enzymes in Rodent Pleural Mesothelial Cells, Tracheal Epithelial Cells and Lung Fibroblasts After Exposure to Hydrogen Peroxide *In Vitro*. *Am. Rev. Respir. Dis.* 147 (1993) A439.
[225] G. Vile and R. Tyrrell, Oxidative Stress Resulting from Ultraviolet A Irradiation of Human Skin Fibroblasts Leads to a Heme Oxygenase-dependent Increase in Ferritin, *J. Biol. Chem.* 268 (1993) 14678- 1468.
[226] M. Reilly *et al.*, Modulation of Oxidative Stress *In Vivo* In Chronic Cigarette Smokers, *Circulation* 94 (1996) 19-25.
[227] A. Kamal *et al.*, Plasma Lipid Peroxide and Blood Superoxide Dismutase Among Asbestos Exposed workers, *Am. J. Indust. Med.* 21(31) (1992) 341-352).
[228] S. Venkat *et al.*, Radioadaptive response in human lymphocytes in vitro. *J. Environ. Pathol. Toxicol. Oncol.* 20(3) (2001) 165-75.
[229] **A. Niess** *et al.*, Free Radicals and Oxidative Stress in Exercise--Immunological Aspects, *Exerc. Immunol. Rev.* 5 (1999) 22-56.
[230] .C. Cobbs *et al.*, Manganese Superoxide Dismutase Expression in Human Central Nervous System Tumors, *Cancer Res.* 56(14) (1996) 3192-3195.
[231] M. Malafa *et al.*, MnSOD Expression Is Increased in Metastatic Gastric Cancer, *J. Surg. Res.* 88(2) (2000) 130-4.
[232] Lee *et al.*, Induction of Apoptosis in p53-deficient Human Hepatoma Cell Line by Wild-type p53 Gene Transduction: Inhibition by Antioxidant, Mol. Cells 12(1) (2001) 17-24.
[233] B. Wedi *et al.*, Eosinophil Apoptosis Is Mediated by Stimulators of Cellular Oxidative Metabolisms and Inhibited by Antioxidants: Involvement of a Thiol-Sensitive Redox Regulation in Eosinophil Cell Death, *Blood* 94(7) (1999) 2365-73.
[234] A. Fornace *et al.*, Ubiquitin is a Major Stress-inducible Transcript In Mammalian Cells, *Nucleic Acid Res.* 17 (1989) 1215-1230.
[235] M. Schlesinger, Heat Shock Proteins. *J. Biol. Chem.* 265 (1990) 12111-12114.
[236] S. Dias *et al.*, VEGF(165) promotes survival of leukemic cells by Hsp90-mediated induction of Bcl-2 expression and apoptosis inhibition. *Blood* 99(7) (2002) 2532-40.
[237] S. Lee *et al.*, Role of Inducible Heat Shock Protein 70 in Radiation-Induced Cell Death, *Cell Stress Chaperones* 6(3) (2001) 273-81.
[238] W. Pratt, The Hsp90-based Chaperone System: Involvement in Signal Transduction From a Variety of Hormone and Growth Factor Receptors, *Proc. Soc. Exp. Biol. Med.* 217(4) (1998) 420-34.
[239] M. Cardillo *et al.*, Heat shock Protein-90, IL-6 and IL-10 in Bladder Cancer, *Anticancer Res.* 20(6B) (2000) 4579-83.
[240] M. Kiriyama *et al.*, Expression of a Small Heat Shock Protein 27 (HSP27) in Mouse Skin Tumors Induced by UVB-irradiation, *Biol. Pharm. Bull.* 24(2) (2001) 197-200.
[241] T. Noguchi *et al.*, Expression of Heat Shock Protein 70 In Grossly Resected Esophageal Squamous Cell Carcinoma, *Ann. Thorac. Surg.* 74(1) (2002) 222-6.
[242] S. Venkat *et al.*, Radioadaptive Response In Human Lymphocytes *In Vitro*, *J. Environ. Pathol. Toxicol. Oncol.* 2001;20(3) (2001) 165-75.
[243] C. Timblin *et al.*, GRP78, HSP72/73, and cJun Stress Protein Levels In Lung Epithelial Cells Exposed to Asbestos, Cadmium, or H2O2, *Free Radic. Biol. Med.* 24(4) (1998) 632-42.
[244] Baker and P. Board, The Human Ubiquitin Gene Family: Structure of a Gene and Pseudogenes from the UbB Subfamily, *Nucleic Acid Res.* 15 (1987) 443-463.
[245] O. Wiborg *et al.*, The Human Ubiquitin Multi-gene Family: Some Genes Contain Multiple Directly Repeated Ubiquitin Coding Sequences, *EMBO J.* 3 (1595) 755-759.
[246] E. Mimnaugh *et al.*, Rapid Deubiquitination of Nucleosomal Histones in Human Tumor Cells Caused by Proteasome Inhibitors and Stress Response Inducers: Effects on Replication, Transcription, Translation, and the Cellular Stress Response, *Biochemistry* 36(47) (1997) 14418-29.
[247] D. Finley *et al.*, The Tails of Ubiquitin Precursors are Ribosomal Proteins Whose Fusion to Ubiquitin Facilitates Ribosome Biogenesis, *Nature* 338 (1985) 394-401.
[248] J. Sonneborn and S. Barbee, Exercise-induced Stress Response as an Adaptive Tolerance Strategy, *Environ. Health Perspect.* 106(Suppl 1) (1998) 325-330.
[249] J. Delic *et al.*, Ubiquitin Pathway Involvement in Human Lymphocyte Gamma-Irradiation-Induced Apoptosis, *Mol. Cell Biol.* 13(8) (1993) 4875-83.

[250] N. Li and M. Karin, Ionizing Radiation and Short Wavelength UV Activate NF-κB Through Two Distinct Mechanisms, *Cell Biology* 95(22) (1998) 13012-13017.
[251] W. Meikrantz and R. Schlegel, Apoptosis and the Cell Cycle, *J. Cell Biochem.* 58 (1995) 160-174.
[252] T. McDonnell, Cell Division Versus Cell Death: A Functional Model of Multistep Neoplasia, Mol. Carcinog. 8 (1993) 209-213.
[253] R. Schulte-Hermann *et al.*, Tumor Development and Apoptosis, *Int. Arch. Allergy Immunol.* 105 (1994) 363-367.
[254] S. James *et al.*, Upregulation of Apoptosis with Dietary Restriction: Implications for Carcinogenesis and Aging *Environ. Health Perspect.* 106(Suppl 1) (1998) 307-312.
[255] S. Phaneuf and C. Leeuwenburgh, Apoptosis and Exercise, *Med. Sci. Sports Exerc.* 33(3) (2991) 393-6.
[256] M. Bijil *et al.*, Effect of Smoking on Activation Markers, Fas Expression and Apoptosis of Peripheral Blood Lymphocytes, *Eur. J. Clin. Invest.* 31(6) (2001) 550-553.
[257] R. Wood *et al.*, Human DNA Genes, *Science 291*(5507) (2001) 1284.
[258] L. Loeb, Microsatellite Instability: Marker of a Mutator Phenotype in Cancer, *Cancer Res.* 54 (1994) 5059-5063.
[259] P. Modrich, Mismatch Repair, Genetic Stability and Cancer, *Science* 266 (1994) 1959-1960.
[260] D. Sagher *et al.*, Heterogeneity of O^6-alkylguanine-DNA Alkyltransferase Activity in Peripheral Blood Lymphocytes: Relationship Between This Activity in Lymphocytes and in Lymphoblastoid Lines from Normal Controls and from Patients with Hodgkin's disease or Non-Hodgkin's Lymphoma. *Cancer Res.* 49 (1989) 5339-5344.
[261] F. Oesch and S. Klein S. Relevance of Environmental Alkylating Agents to Repair O^6-alkylguanine-DNA Alkyltransferase: Determination of Individual and Collective Repair Capacities of O^6-methylguanine. *Cancer Res.* 52 (1992) 1801-1803.
[262] J. Madden *et al.*, Effects of Opiates and Demographic Factors on DNA Repair Synthesis in Human Leukocytes, *Proc. Natl. Acad. Sci. USA* 76 (1979) 5769-5773.
[263] Celotti *et al.*, DNA Repair and Replication in Lymphocytes from Smokers Exposed *In Vitro* to UV-Light. *Mutagenesis* 4 (1989) 82-86.
[264] J. Mayer *et al.*, Biologic Markers in Ethylene Oxide-Exposed Workers and Controls, *Mutat Res* 248 (1991) 163-176.
[265] K. Helzlsouer *et al.*, DNA Repair Proficiency: Potential Susceptibility Factor for Breast Cancer, *J. Natl. Cancer Inst.* 88 (1996) 754-755.
[266] Q. Wei Q *et al.*, Reduced DNA Repair Capacity in Lung Cancer Patients. *Cancer Res.* 56 (1996) 4103-4107.
[267] A. Auerbach and R. Allen, Leukemia and Preleukemia in Fanconi Anemia Patients: A Review of the Literature and Report of the International Fanconi Anemia Registry, *Cancer Genet. Cytogenet.* 51 (1991) 1-12.
[268] A. Taylor *et al.*, Leukemia and Lymphoma in Ataxia Telangiectasia, *Blood* 87 (1996) 423-438.
[269] D. Bell *et al.*, Genetic Monitoring of Human Polymorphic Cancer Susceptibility Genes by Polymerase Chain Reaction: application to glutathione transferase μ, *Environ. Health Perspect.*, 98 (1992) 113–117.
[270] W. Au *et al.*, ‚Monitoring Populations for DNA Repair Deficiency and for Cancer Susceptibility, *Environ. Health Perspect.*, 104(Suppl 3):579-584 (1996)
[271] T. Nakajima *et. al.*, Expression and Polymorphism of Glutathione *S*-Transferase in Human Lungs: Risk Factors in Smoking-Related Lung Cancer. *Carcinogenesis*, **16** (1995) 707–711.
[272] S. Kato *et. al.*, (1992) Cytochrome P450IIE1 Genetic Polymorphisms, Racial Variation and Lung Cancer. *Cancer Res.*, **52** (1992) 6712–6715.
[273] T. Oyama *et. al.*, Cytochrome P450 2E1 Polymorphism as a Risk Factor for Lung Cancer: In Relation to p53 Gene Mutation. *Anticancer Res.*, **17** (1997) 583–588.
[274] S. Hayashi *et. al.*, (1991) Genetic Linkage of Lung Cancer-Associated MspI Polymorphisms with Amino Acid Replacement in the Heme Binding Eegion of the Human Cytochrome P450IA1 Gene. *J. Biochem.*, 110 (1991) 407–411.
[275] M. You *et al.*, Activation of the Ki-*ras* Protooncogene in Spontaneously Occurring and Chemically Induced lung Tumors of the Strain A mouse, *Proc. Natl. Acad. Sci. USA* 86 (1989) 3070-3074.
[276] M. Greenblatt *et al.*, Mutations in the *p53* Tumor Suppressor Gene: Clues to Cancer Etiology and Molecular Pathogenesis, *Cancer Res.* 54 91994) 4855-4878.
[277] D. Malkin Age-specific Oncogenesis: The Genetics of Cancer Susceptibility Environ Health Perspect 103(Suppl 5) (1995) 45-48.
[278] M. Verma and S. Srivastava, Epigenetics in Cancer: Implications for Early Detection and Prevention, *Lancet Oncol.* 3(12) (2002) 755-63.

Toxicogenomics and Proteomics
J.J. Valdes and J.W. Sekowski (Eds.)
IOS Press, 2004

Expression Profiling of Sulfur Mustard Exposure in Murine Skin: Chemokines, Cytokines, and Growth Factors

Carol L.K. SABOURIN[1], Young W. CHOI[1], Mindy K. STONEROCK[1], Jack D. WAUGH[1], Robyn C. KISER[1], Michele M. DANNE[1], Kristi L. BUXTON[1], Robert P. CASILLAS[1], Michael C. BABIN[1], and John J. SCHLAGER[2]

[1]Battelle Memorial Institute, Medical Research and Evaluation Facility, Columbus, OH, USA; [2]US Army Medical Research Institute of Chemical Defense, Pharmacology Division, Aberdeen Proving Ground, MD, USA

Abstract. The chemical warfare agent sulfur mustard (SM) produces a delayed inflammatory response followed by blister formation in skin of exposed individuals. Severity and time to onset of inflammation and tissue damage vary greatly and depend on various factors such as dose, ambient temperature, and exposure site. Gene arrays provide the opportunity to identify multiple transcriptional events associated with low-level exposure SM. This study examined SM-induced changes in gene expression in skin from mice cutaneously exposed to SM using cDNA arrays. Skin from 5 mice, paired as SM-exposed right ear and unexposed left ear at six dose levels (0.005, 0.01, 0.02, 0.04, 0.08, and 0.16 mg) was harvested at 24 hours post-exposure. The lowest concentration (0.005 mg) produced no apparent tissue damage. The highest SM dose (0.16 mg) produced severe injury characterized by edema, dermal infiltration of inflammatory cells, premature death of basal layer epidermal cells, and epidermal-dermal separation. We analyzed SM-induced genes using cDNA microarray technology, which included 1,176 genes. The genes were classified in six categories. Chemokines, cytokines and growth factors represented a large number of the genes altered with SM exposure. A number of these genes not known to be associated with SM toxicity were identified. The results provide not only a new molecular basis for understanding the response to SM, but also a useful resource for future development of diagnostic markers and therapeutic targets for SM injury.

1. Introduction

Sulfur mustard [bis-(2-chloroethyl)-sulfide] (SM) is a chemical warfare agent for which there is no defensive pharmacological treatment for either use prior to or following exposure. SM is easily prepared, a true threat to the warfighter, and as noted by terrorist propaganda, a weapon that may be used on citizens. SM is an alkylating agent with cytotoxic, mutagenic, and vesicating properties [1]. Cutaneous exposure to SM produces a delayed inflammatory skin response and severe tissue injury. Onset and severity of skin injury is dependent on dose, skin moisture, body site, and ambient temperature. Erythema appears within a few hours of exposure followed by edema and blister formation.

Histopathologically, cutaneous exposure to SM in animal or man is characterized by edema, dermal infiltration of inflammatory cells, premature death of basal layer epidermal cells, and epidermal-dermal separation [2-11]. The biochemical mechanism for SM-induced skin injury is not well understood.

We have used the mouse ear vesicant model (MEVM) to evaluate medical countermeasures against SM and to elucidate the molecular events that follow cutaneous exposure to SM [12,13]. In this model a liquid challenge consisting of 5 μ L of 195 mM SM in dichloromethane permitted flash surface dispersion of SM when applied to the inner surface of the right ear of mice. Therapeutic compounds were delivered topically to the same SM-treated ear region at specified pre- and post- treatment times. Over 400 compounds have been screened for protective efficacy against SM using the MEVM. Compounds were selected from five defined therapeutic strategies: anti-inflammatory, anti-protease, scavenger/chelator, deoxyribonucleic acid (DNA) repair, and cell-cycle regulator compound classes. Lead compounds have been identified that protect against SM tissue damage as measured by two or more indicators used to evaluate injury (edema, subepidermal blister, tissue necrosis) and demonstrated a >50% reduction in at least one endpoint. These compounds comprised three protection strategies: anti-inflammatory, scavenger, and anti-protease [13].

Recent studies have identified specific gene activation of inflammatory mediators and proteases suggesting a role for these proteins in SM-induced skin injury. Our molecular studies have identified biomarkers of SM injury for use in assessing effective candidate pretreatments for SM and to establish molecular signatures for SM exposure [14-16]. Early increases in inflammatory gene expression of cytokines were identified using quantitative RT-PCR and immunohistochemistry in murine and porcine models. This study utilized microarrays to identify novel markers and mechanisms of action through analysis of the gene expression patterns resulting from exposure to SM. The identification of these genes using a DNA array approach will aid in identifying potential candidates for pharmacological intervention. These studies provide a molecular basis for understanding the toxicity of SM and provide resources for future development of therapeutic and diagnostic markers.

2. Material and Methods

2.1 SM Cutaneous Exposure

Male CD1 mice, approximately 4 weeks of age were purchased from Charles River Laboratories (Portage, MI) and housed in a controlled environment with a 12 h light/dark cycle. Purina Certified Rodent Chow® and water was available *ad libitum*. On the study day, mice weighing in the range of 25 to 35 g were marked for identification and anesthetized with a combination of ketamine (60 mg/kg) and xylazine (12 mg/kg) by intraperitoneal injection. A single 5 μ L application of various SM solutions in dichloromethane was applied to the inner surface of the right ear in a fume hood according to published procedures (12). Decreasing concentrations from 195 mM through 6 mM SM were prepared permitting application of SM doses of 0.16, 0.08, 0.04, 0.02, 0.01, and 0.005 mg in 5 μL. The solution of SM allowed even distribution of agent over the entire medial surface of the ear. The left ear (vehicle control) was exposed to 5 μL of dichloromethane only. Animals were housed two per cage after SM challenge and cages were placed on warm water-perfused heating pads within the laboratory fume hood system. At 24 h post SM challenge, animals were euthanized in a halothane-filled chamber. Ears were collected from 5 animals and immediately frozen in liquid nitrogen for comparison of gene expression.

2.2 RNA Isolation

Total RNA was isolated from frozen ear tissue using a standard phenol/chloroform extraction method that employed RNAzol B reagent (Tel-Test, Inc., Friendswood, TX) according to the manufacturer's recommendations. Total RNA was quantified by measuring the absorbance at 260 nm using a spectrophotometer. The RNA was evaluated for any degradation or impurities by calculating the ratio of OD_{260} to OD_{280} and by electrophoresis in a denaturing agarose gel.

2.3 Analysis of mRNA Expression Using cDNA Microarrays

Preparation of radioactively labeled cDNA and subsequent hybridizations to the Atlas Mouse 1.2 cDNA microarrays was conducted as recommended by the manufacturer using reagents provided by the manufacturer (BD Biosciences Clontech, Palo Alto, CA.). Briefly, total RNA was DNase-treated using an organic-free extraction method (Ambion, Austin, TX). Poly A+ RNA was then reverse-transcribed in the presence of 50 μCi of [α-^{33}P]dATP. The labeled cDNA was purified by column chromatography and incorporation of the ^{33}P radionucleotide was determined by liquid scintillation counting. The arrays were prehybridized in ExpressHyb solution (BD Biosciences Clontech) containing 100 μg/mL of denatured salmon sperm DNA for 30 minutes at 68°C. Cot-1 DNA was used for blocking of unspecific hybridization (BD Biosciences Clontech). The probe was denatured at 95-100°C for 2 minutes, and then placed on ice for 2 minutes. Final probe concentration was equivalent for the treated and untreated samples and at minimum 10^6 cpm. Hybridization to the Atlas membrane was carried out overnight at 68°C. The membrane was washed three times in 2X SSC, 1% SDS at 68°C for 30 min and twice in 0.1X SSC, 0.5% SDS at 68°C for 30 min. To identify positive hybridization signals, the membranes were exposed to a phosphor screen (Molecular Dynamics, Sunnyvale, CA) for approximately 3 to 4 days, and then visualized with a phosphorimager (Molecular Dynamics).

Hybridized dot intensities on the microarrays were quantified using AtlasImage 2.0 software (BD Biosciences Clontech). Normalization of the data was performed by subtracting the global background value of each microarray calculated by AtlasImage from the intensity value for each gene and then dividing the background-subtracted intensity value for each gene by the average intensity of all of the genes on that microarray. Average normalized intensity values were calculated for all genes derived from each microarray. Ratio values were calculated by dividing the average normalized intensity value for a gene from SM-exposed ear by the average normalized intensity value of that same gene from the matched control ear. Further data analysis was performed using Atlas Navigator 2.0 software (BD Biosciences Clontech). Genes were selected on the basis of 2 of 5 mice in the same dose group demonstrating an increase or decrease in gene expression of ≥2-fold for the SM-exposed compared to the control ear, and an alteration at two or more doses of SM.

3. Results and Discussion

The mouse ear inflammation model was developed to establish simple endpoints of skin injury following cutaneous exposure to SM. Mouse ear edema and histopathologic response to a single topical application of 0.16 mg of SM is evaluated at 24 h postexposure [10, 12]. Histopathological features of SM cutaneous exposure include the formation of microscopic blisters or vesicles at the epidermal-dermal junction (subepidermal blister), and epidermal necrosis (Figure 1). In contrast, these changes are not observed at 0.005 mg SM, the lowest dose utilized in this study, and the skin resembles that exposed to the vehicle, dichloromethane (data not shown).

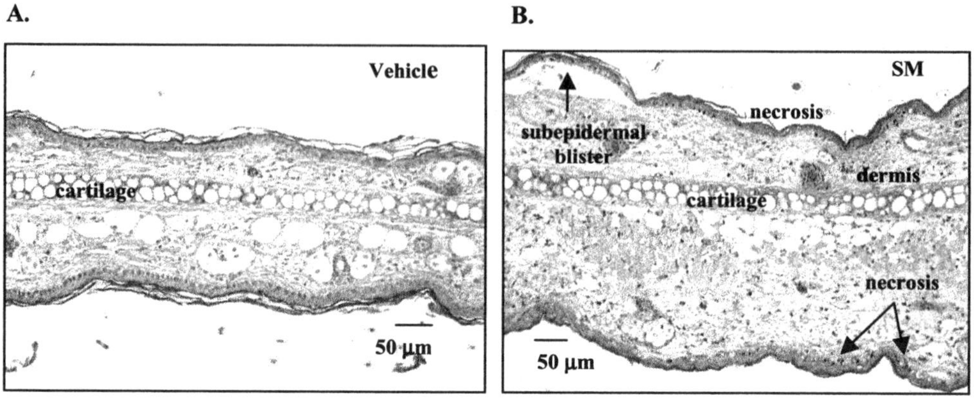

Figure 1. Light microscopy of H&E stained mouse ear skin 24 h after topical exposure to (A) dichloromethane, vehicle control and (B) 0.16 mg of SM.

In this study, SM-exposed and control ear skin from five animals at each of six doses (0.005, 0.01, 0.02, 0.04, 0.08, and 0.16 mg) of SM were analyzed using Mouse Atlas 1.2 cDNA microarrays consisting of 1,176 genes per array. Relative expression was determined for 5 individual animals at each of 6 doses of SM. The criteria for gene transcript alteration criteria were a $\geq$2-fold increase for 2 of 5 mice in the same dose group, and an alteration at two or more doses of SM. Chemokines, cytokines, and growth factors showed a large number of altered genes in the SM-exposed skin. Altered genes in skin as a result of in vivo exposure to SM at 24 h are presented in Table 1. Thirty genes were up-regulated and 18 genes were down-regulated as a result of exposure to SM at two or more doses. Interleukin-1β (IL-1β) showed increased gene expression as previously observed using RT-PCR with exposure to 0.16 mg SM at 24 h [15]. However, genes not previously known to be altered by SM were identified.

Table 1. Altered Genes in SM-Exposed Mouse Ear Skin as Determined by Microarray Analysis

Alteration	Gene Name	Accession
Up-Regulated	C5A receptor	L05630
	CD28 antigen	M34563
	chemokine (C-C) receptor 1	U29678
	chemokine (C-C) receptor 2	U56819
	chemokine (C-X-C) receptor 2	D17630
	colony stimulating factor 3 (granulocyte)	M13926
	colony stimulating factor 3 receptor (granulocyte)	M58288
	Follistatin	Z29532
	growth differentiation factor 9	X77113
	heparin binding epidermal growth factor-like growth factor	L07264
	insulin-like growth factor I receptor	U00182
	interleukin 1 beta	M15131
	interleukin 4 receptor, alpha	M27959
	Interleukin 5 receptor alpha	D90205
	interleukin 7	X07962
	interleukin 9 receptor	M84746
	kit oncogene	Y00864
	oncostatin M	D31942
	platelet derived growth factor receptor, alpha polypeptide	M84607
	macrophage inflammatory protein 1 alpha	X12531
	macrophage inflammatory protein 1 beta	M23503
	monocyte chemoattractant protein 3	S71251
	nerve growth factor, beta	K01759
	small inducible cytokine subfamily, member 2	X53798
	tumor necrosis factor (ligand) superfamily, member 5	X65453
	tumor necrosis factor (ligand) superfamily, member 6	U06948
	tumor necrosis factor receptor superfamily, member 1b	M59378
	tumor necrosis factor receptor superfamily, member 6	M83649
	tumor necrosis factor receptor superfamily, member 7	L24495
	tumor necrosis factor receptor superfamily, member 8	U25416
Down-Regulated	activin A receptor, type 1B	Z31663
	bone morphogenetic protein 8a	M97017
	endothelin 3	U32330
	Eph receptor B2	L25890
	epidermal growth factor	J00380
	erythropoietin receptor	S59388
	fibroblast growth factor receptor 1	M28998
	FMS-like tyrosine kinase 3 ligand	U04807
	inhibin alpha	X69618
	insulin-like growth factor binding protein 3	X81581
	insulin-like growth factor binding protein 5	X81583
	interferon gamma receptor	M28233
	interleukin 11	U03421
	interleukin 4	M25892
	interleukin 6 receptor, alpha	X51975
	kinase insert domain protein receptor	X70842
	platelet derived growth factor receptor, beta polypeptide	X04367
	transforming growth factor, beta 2	X57413

The gene expression for monocyte chemoattractant protein 3 (MCP-3) was increased at multiple doses of SM, whereas macrophage inflammatory protein 1 α (MIP-1α) and MIP-1β showed increased expression only at the two highest doses of SM, 0.08 and 0.16 mg (Figure 2). Bone morphogenetic protein 4 and insulin-like growth factor binding protein 3 showed decreased expression at doses of 0.02 mg SM and higher (Figure 3). Interleukin (IL-4) receptor α expression was increased, whereas IL-4 was decreased at most doses of SM (Figure 4). Six members of the tumor necrosis factor (TNF) superfamily showed increased expression as a result of SM-exposure, including both TNF ligands and receptors. Gene expression was altered in proteins that play a role in the regulation of the immune system.

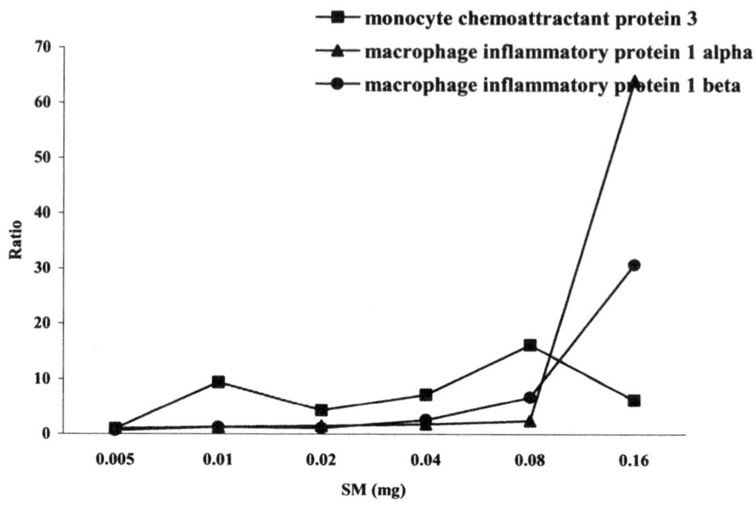

Figure 2. Gene expression for MCP-3, MIP-1α, and MIP-1β in skin at 24 h post-exposure to SM. The mean ratio of the level for the SM-exposed compared to the control ear was plotted for five animals at each dose of SM.

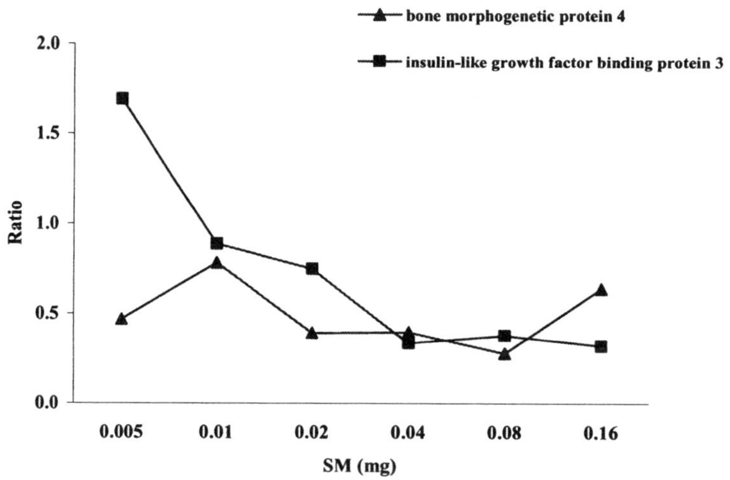

Figure 3. Gene expression for bone morphogenetic protein 4 and insulin-like growth factor binding protein 3 in skin at 24 h post-exposure to SM. The mean ratio of the level for the SM-exposed compared to the control ear was plotted for five animals at each dose of SM.

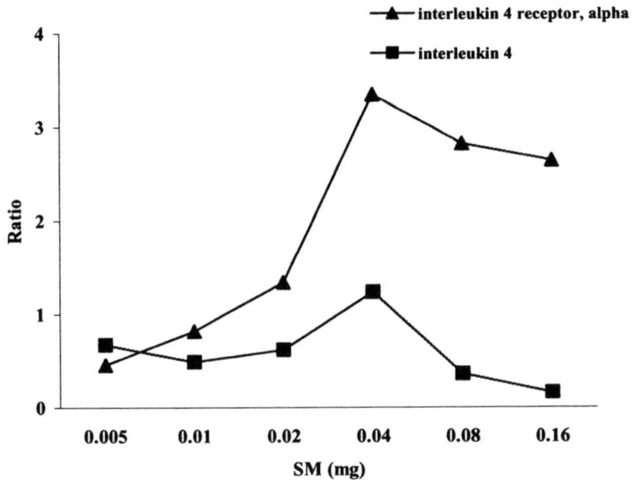

Figure 4. IL-4 and the IL-4 receptor α gene expression in skin at 24 h post-exposure to SM. The mean ratio of the level for the SM-exposed compared to the control ear was plotted for five animals at each dose of SM.

Molecular profiling using microarrays provides valuable insights into alterations in gene expression. The identification of these genes by microarray may indicate new targets for intervention in the inflammatory response to protect against SM-induced injury and improve healing time.

Acknowledgments

This work was supported by the U.S. Army Medical Research and Materiel Command under Contract No. DAMD17-99-D-0010. The view, opinions and/or findings contained in this report are those the author(s) and should not be construed as an official Department of the Army position, policy or decision unless so designated by other documentation. In conducting research using animals, the investigator(s) adhered to the *Guide for the Care and Use of Laboratory Animals* prepared by the Institute of Laboratory Animal Resources, National Research Council (National Academy Press, 1996).

References

[1] F. R. Sidell, J.S. Urbanetti, W.J. Smith, and C.G. Hurst, Vesicants. In: F.R. Sidell, E.T. Takafuji, and D.R. Franz (editors). Textbook of Military Medicine, Warfare, Weaponry, and the Casualty. Medical Aspects of Chemical and Biological Warfare. TMM Publications, Washington D.C., 1997, pp 197-228.

[2] B. Papirmeister, C.L. Gross, J.P. Petrali, and C.J. Hixson, Pathology Produced by Sulfur Mustard in Human Skin Grafts on Athymic Nude Mice I. Gross and Light Microscopic Changes, *J. Toxicol. Cutan. Ocul. Toxicol.* **3** (1984) 371-408.

[3] L.W. Mitcheltree, M.M. Mershon, H.G. Wall, and J.D. Pulliam, Microblister Formation in Vesicant-Exposed Pig Skin, *J. Toxicol. Cutan. Ocul. Toxicol.* **8** (1989) 309-319.

[4] J.P. Petrali, S.B. Oglesby, and K.R. Mills, Ultrastructural Correlates of Sulfur Mustard Toxicity, *J. Toxicol. Cutan. Ocul. Toxicol.* **9** (1990) 193-214.

[5] M.M. Mershon, L.W. Mitcheltree, J.P. Petrali, E.H. Braue, and J.V. Wade, Hairless Guinea Pigs Bioassay Model for Vesicant Vapor Exposures, *Fund. Appl. Toxicol.* **15** (1990) 622-630.

[6] V.P. Barranco, Mustard Gas and the Dermatologist, *Int. J. Dermatol* **30** (1991) 684-686.

[7] W.J. Smith and M.A. Dunn, Medical Defense Against Blistering Chemical Warfare Agents, *Arch. Dermatol.* **27** (1991) 1207-1213.
[8] J. Borak and F.R. Sidell, Agents of Chemical Warfare: Sulphur Mustard, *Anal. Emerg. Med.* **21** (1992) 303-308.
[9] K.J. Smith, H.G. Skelton, D.W. Hobson, F.M. Reid, J.A. Blank, and C.G. Hurst, Cutaneous Histopathologic Features in Weanling Pigs After Exposure to Three Different Doses of Liquid Sulfur Mustard, *Am. J. Dermatopathol.* **18** (1996) 515-520.
[10]K.J. Smith, R. Casillas, J. Graham, H.G. Skelton, F. Stemler, and B.E. Hackley Jr., Histopathologic Features Seen with Different Animal Models Following Cutaneous Sulfur Mustard Exposure, *J. Dermatol. Sci.* **14** (1997) 126-135.
[11]K.J. Smith, W.J. Smith, T. Hamilton, H.G. Skelton, J.S. Graham, C. Okerberg, R. Moeller, and B.E. Hackley Jr., Histopathologic and Immunohistochemical Features in Human Skin After Exposure to Nitrogen and Sulfur Mustard, *Am. J. Dermatopathol.* **20** (1998) 22-28.
[12]R.P. Casillas, L.W. Mitcheltree, and F.W. Stemler, The Mouse Ear Model of Cutaneous Sulfur Mustard Injury, *Toxicol. Methods* **7** (1997) 381-397.
[13]R.P. Casillas, R.C. Kiser, J.A. Truxall, A.W. Singer, S.M. Shumaker, N.A. Niemuth, K.M. Ricketts, L.W. Mitcheltree, L.R. Castrejon, and J.A. Blank, Therapeutic Approaches to Dermatotoxicity by Sulfur Mustard. I. Modulation of Sulfur Mustard-Induced Cutaneous Injury in the Mouse Ear Vesicant Model (MEVM), *J. Appl. Toxicol.* **20** (2000) S145-S151.
[14]K.M. Ricketts, C.T. Santai, J.A. France, A.M. Graziosi, T.D. Doyel, M.Y. Gazaway, and R.P. Casillas. Inflammatory Cytokine Response in Sulfur Mustard-Exposed Mouse Skin, *J. Appl. Toxicol.* **20** (2000) S73-S76.
[15]C.L.K. Sabourin, J.P. Petrali, and R.P. Casillas. Alterations in Pro-Inflammatory Mediator Gene Expression in Sulfur Mustard Exposed Mouse Skin, *J. Biochem. Mol. Toxicol.* **14** (2000) 291-302.
[16]C.L.K. Sabourin, M.M. Danne, K.L. Buxton, R.P. Casillas, and J.J. Schlager. Cytokine, Chemokine, and Matrix Metalloproteinase Response after Sulfur Mustard Injury to Weanling Pig Skin, *J. Biochem Mol Toxicol* **16**, (2003) 263-272.

Toxicogenomics and Proteomics
J.J. Valdes and J.W. Sekowski (Eds.)
IOS Press, 2004

Further Progress in DNA Repair Puzzle in the Postgenomic Era

Janusz KOCIK
Military Institute of Hygiene and Epidemiology, Warsaw, Poland

Abstract. With the completion of the genome sequence of the consecutive Prokaryota and Eukaryota these organisms can be analyzed on the genome scale. With DNA microarray technology it is now possible to measure the mRNA levels of thousands of cellular genes under variety of experimental conditions. The advent of ultrasensitive spectrometric protein identification methods allows to directly identify protein complexes on a proteome-wide scale. The approaches to decipher protein-protein interactions are being shifted from detection of binary associations to revealing of network associations through more holistic and fine-tuned methods. Herein the opportunities for further progress in DNA repair revealing are shown with a special attention drawn to different model organisms, including that ones which are very resistant to environmental conditions and in which DNA repair processes are scrupulously studied.

1. Introduction

Since most mutations are deleterious, no species can afford to allow them to accumulate at a high rate in its germ cells. Mutation frequency is thought to limit the number of essential proteins that any organism can encode in its germ line to about 60,000. By an extension a mutation frequency tenfold higher would limit an organism to about 6000 essential proteins. In this case evolution would probably have stopped at an organism no more complex than a fruit fly. Therefore, for DNA to serve as the genetic link between generations, the base sequence must be not only copied correctly during replication, but also maintained throughout the life span of cells [5].

Fewer than one in a thousand accidental base changes in DNA causes a mutation; the rest are eliminated with remarkable efficiency by DNA repair. There are a variety of repair mechanisms, each catalyzed by a different set of enzymes. Nearly all of these mechanisms depend on the existence of two copies of the genetic information, one in each strand of the DNA double helix.

During DNA replication mispaired bases are instantly excised by polymerase. Additionally many post replication repair systems exists that can be roughly divided into three broad categories:

- Mismatch repair, which occurs immediately after DNA synthesis, uses the parental strand as a template to correct an incorrect nucleotide incorporated into the newly synthesized strand.
- Excision repair that entails removal of a damaged region by specialized nuclease systems and then DNA synthesis to fill the gap.
- Repair of double-strand DNA breaks in multicellular organisms that occurs primarily by an end-joining process.

2. Proofreading by DNA Polymerase

The increased accuracy of replication is largely due to the proofreading function of DNA polymerases. In DNA polymerase III, this function resides in the ε subunit of the core polymerase. When an incorrect base is incorporated during DNA synthesis, the polymerase pauses, then transfers the 3′ end of the growing chain to the exonuclease site where the mispaired base is removed. Then the 3′ end is transferred back to the polymerase site, where this region is copied correctly. Proofreading is a property of almost all bacterial DNA polymerases. Both the δ and ε DNA polymerases of animal cells, but not the α polymerase, also have proofreading activity. It seems likely that this function is indispensable for all cells to avoid excessive genetic damage.

3. Mismatch Repair

Bacterial and eukaryotic cells have a mismatch-repair system that recognizes and repairs all single-base mispairs, as well as small insertions and deletions. The problem with mismatch repair is determining which is the normal and which is the mutant DNA strand, and repairing the latter so that it is properly base-paired with the normal strand. How this is accomplished has been elucidated for the *E. coli* methyldirected mismatch-repair system, often referred to as the MutHLS system.

In *E. coli* DNA, adenine residues in a GATC sequence are methylated at the 6 position only on the parental strand. The adenines in GATC sequences on the daughter strands are methylated by a specific enzyme, called Dam methyltransferase, only after a lag of several minutes. During this lag period, the newly replicated DNA contains hemimethylated GATC sequences. An E. coli protein designated MutH, which binds specifically to hemimethylated sequences, is able to distinguish the methylated parental strand from the unmethylated daughter strand. If an error occurs during DNA replication, resulting in a mismatched base pair near a GATC sequence, another protein, MutS, binds to this abnormally paired segment, triggers binding of MutL, that connects MutS with a nearby MutH. This event activates a latent endonuclease activity of MutH, which then cleaves specifically the unmethylated daughter strand. The segment of the daughter strand containing the misplaced base is excised and replaced with the correct DNA sequence.

A similar mechanism repairs lesions resulting from depurination, the loss of a guanine or adenine base from DNA. Apurinic sites generate mutations during DNA replication. Apurinic endonucleases cut a DNA strand near an apurinic site. As with mismatch repair, the cut is extended by exonucleases, and the resulting gap then is repaired by DNA polymerase and ligase.

In human the mismatch is recognized by MutS homologs, perhaps most often MSH2 and GTBP/MSH6, though another MutS homolog, MSH3, may substitute for GTBP/MSH6 in some cases. MutL homologs, such as MLH1 and PMS2, are recruited to the complex and the mismatch is repaired through the action of a number of proteins, including an exonuclease, helicase, DNA polymerase, and ligase.

4. Base Excision Repair

Base excision repair involves a battery of enzymes called DNA glycosylases. Each DNA glycosylase recognizes an altered base in DNA and catalyzes its hydrolytic removal. The DNA glycosylase reaction produces a deoxyribose sugar with a missing base recognized by the AP endonuclease the subsequent steps in the repair process proceed in the same way as for depurinated sites. Cells use excision repair to fix DNA regions containing chemically modified bases, called chemical adducts, that distort the normal shape of DNA locally. A key to this type of repair is the ability of certain proteins to slide along the surface of a double-stranded DNA molecule looking for bulges or other irregularities in the shape of the double helix.

The best-understood example of excision repair is the UvrABC system from *E. coli.* Cells carrying mutations in the uvrA, B, or C locus are very sensitive to UV light and chemicals that add large groups to DNA. The UvrA-UvrB complex initially binds to an undamaged segment and translocates along the DNA helix until a distortion caused by an adduct is recognized. Conformational change in the damaged DNA region bound to the UvrA-UvrB complex produces a bend, or kink, in the DNA backbone. After the UvrA dimer has dissociated, the UvrC protein, which has endonuclease activity, binds to the damaged site. The interaction of UvrC and the bent DNA is thought to open up space within the DNA, allowing the enzyme to access its target. After UvrC has cleaved the damaged strand at two points, the fragment with the adduct is removed by a helicase and degraded; the gap left in the strand then is repaired by the combined actions of DNA polymerase I and DNA ligase.

5. End-Joining Repair of Nonhomologous DNA

Double-strand breaks are caused by ionizing radiation and by anticancer drugs, such as bleomycin, which is why these drugs are used to kill rapidly growing cells. A cell that has suffered a particular double-strand break usually contains other breaks; such breaks can be repaired by joining the free DNA ends. The joining of broken ends from different chromosomes, however, leads to translocation of pieces of DNA from one chromosome to another. Such translocations may trigger abnormal cell growth by placing a proto-oncogene under the inappropriate control of a promoter from another gene.

Double-strand breaks can be correctly repaired only if the free ends of the DNA rejoin exactly. Such repair is complicated by the absence of single-stranded regions that can direct base-pairing during the joining process. One of the two mechanisms that have evolved to repair double-strand breaks is homologous recombination. In this process the double-strand break on one chromosome is repaired using the information on the homologous, intact chromosome.

In multicellular organisms the predominant mechanism for repairing double-strand breaks involves rejoining the ends of the two DNA molecules. Although this process yields a continuous double-stranded molecule, it results in loss of several base pairs at the joining point. Formation of such a possibly mutagenic deletion is one example of how repair of DNA damage can introduce mutations. A similar process can link together any two DNA molecules, even those cut from different chromosomes.

Genetic studies in eukaryotes ranging from yeast to humans suggest that quite similar excision-repair mechanisms are employed by different organisms. The basic strategy is to search for mutants that exhibit increased sensitivity to UV light or other agents that produce shape-distorting lesions in DNA. Such mutants presumably are deficient in the wild-type excision-repair mechanisms that normally repair damage caused by such agents. In the yeast S. cerevisiae, for instance, numerous UV-sensitive mutants and 10 radiation-sensitive (RAD) genes have been identified. Two of the human genes identified in this way have been shown to be related to the yeast RAD3 and RAD10 genes. The human protein with partial homology to the RAD10 protein also contains a region that is similar to part of E. coli UvrC.

6. Inducible Error-Prone DNA-Repair System

When a cell suffers so much DNA damage over a short time that its repair systems are saturated, it runs the danger of extensively replicating unrepaired lesions, thereby perpetuating mutations. In such situations, both bacterial and animal cells use inducible repair systems. Such systems are not expressed in undamaged cells, but some aspect of the accumulated damage causes their derepression (induction) and expression. This inducible systems are used only as a last resort when error-free mechanisms of repair cannot cope with damage.

7. Bacteria SOS System

One such inducible system is the SOS repair system of bacteria. Because this system generates many errors in the DNA as it repairs lesions, it is referred to as error-prone. The SOS system, which repairs UV-induced damage, differs from the constitutive UvrABC system in that its activity is dependent on RecA protein. RecA also participates in homologous recombination. The errors induced by the SOS system are at the site of lesions, suggesting that the mechanism of repair is insertion of random nucleotides in place of the damaged ones in the DNA. Therefore, most of the mutations produced by treating bacteria with radiation or chemicals are caused by the error-prone SOS repair system, not by the original lesions themselves.

Animal cells also have inducible repair systems, although it is not known whether these are error-prone. The main mechanism for repairing double-strand breaks in eukaryotes clearly is error-prone. Thus, double-strand repair and, perhaps, error-prone inducible repair likely play a role in mutagenesis and therefore in carcinogenesis in animals

8. Cell Cycle Control and DNA Damage

Because radiation- or carcinogen-induced DNA damage must be repaired before the DNA is replicated, cells have sensing mechanisms that react to DNA damage and stop DNA replication. As eukaryotic cells move through the cell cycle, specific sets of genes are transcriptionally activated and inactivated, although transcript levels for the vast majority of genes do not change. Moreover, responses to DNA-damaging agents are known to vary throughout the cell cycle; e.g., G1 cells that experience DNA damage activate a G1/S checkpoint, those in S phase activate an S-phase delay, and those in G2 or M activate a G2/M checkpoint. These mechanisms, involve checkpoint control proteins such as the p53 protein, which acts to stop the cell cycle if DNA is damaged, and thus to suppress production of tumors. Cells that do not express functional p53 protein exhibit high rates of mutation in response to DNA damage, accelerating the formation of tumors.

9. DNA Repair Genomics in Radiation Resistant Bacterium

The bacterium *Deinococcus radiodurans* shows remarkable resistance to a range of damage caused by ionizing radiation, desiccation, UV radiation, oxidizing agents, and electrophilic mutagens. *D. radiodurans* is best known for its extreme resistance to ionizing radiation; not only can it grow continuously in the presence of chronic radiation (6 kilorads/h), but also it can survive acute exposures to gamma radiation exceeding 1,500 kilorads without dying or undergoing induced mutation.

The genome of *D. radiodurans* has already been sequenced. Genes that potentially affect DNA repair and recombination and stress responses were investigated in detail by genomic approach through detailed computational analysis of the genome by Makarova et al.[6]. They used several approaches for deeper protein characterization. In particular, they systematically applied sensitive profile-based methods that included PSI-BLAST, which constructs a position-dependent weight matrix from multiple alignments generated from the BLAST hits above a certain expectation value and allows iterative database searches using the information derived from such a matrix; IMPALA, which searches the matrix against profile databases, and SMART, which uses a Hidden Markov Model algorithm to search a sequence against a multiple-alignment database. By this approach the authors investigated the predicted repair system components of *D. radiodurans* in detail, to detect any possible correlation with its exceptional radioresistant and desiccation-resistant phenotype. Generally, it appears that *Deinococcus* possesses a typical bacterial system for DNA repair and that, commensurate with the genome size, its repair pathways even appear to be less complex and diverse than those of bacteria with larger genomes, such as *E. coli* and *B. subtilis. D. radiodurans* has repair pathways that include excision repair, mismatch repair, and recombinational repair. Generally, no marked error-prone SOS response is observed. However, there have been a few reports consistent with SOS response, where preexposure to low doses of ionizing radiation, UV, or hydrogen peroxide causes a low level of subsequent increased resistance to DNA damage (twofold or less). Since the SOS response is not always mutagenic, the absence of DNA damage-induced mutagenesis observed in *D. radiodurans* cannot be taken as evidence against the existence of the SOS response in this bacterium. At the same time, there are several interesting and unusual aspects of the predicted layout of the repair systems in *Deinococcus* that may be linked to its phenotype. In addition to well-characterized repair proteins, *Deinococcus* encodes several unusual proteins and expanded protein families that are less confidently associated with repair but might contribute to the unusual effectiveness of the repair and recombination systems in this bacterium. For example, the expanded Nudix hydrolase superfamily and the homologs of plant desiccation resistance-associated proteins are likely to contribute to both the extreme radiation and the desiccation resistance of *Deinococcus*. Furthermore, the unexpectedly numerous nucleotide repeats may also play a role in stress response. However, the sobering conclusion from this study is that the fundamental questions underlying the extreme resistance phenotype of *D. radiodurans* remain unanswered. It seems most likely that this phenotype is very complex and is determined collectively by some of the features revealed by genome analysis done by Makarova et al., as well as by many more subtle structural peculiarities of proteins and DNA that are not readily inferred from the sequences, at least not with the current limited collection of genomes available for comparative analysis. The authors expect that a comprehensive understanding of the mechanisms of damage repair in *Deinococcus* will arise from a combination of further comparative genomic analysis and prediction-driven experiments.

10. Genome of Halobacterium Species, the Bacteria Resistant to High Salinity

Halobacterium species are obligatory halophilic microorganisms that have adapted to optimal growth under conditions of extremely high salinity - 10 times that of sea water. They contain a correspondingly high concentration of salts internally and exhibit a variety of unusual and unique molecular characteristics. Wailap Ng et al., sequenced *Halobacterium* sp. NRC-1 using a whole genome shotgun strategy [8]. The sequences were assembled by using the PHREDPHRAP programs. The program GLIMMER was used for gene prediction of the finished genome sequence. Predicted genes were translated and the resulting sequences used to search the nonredundant database of proteins (translation of GenBank CDS, Protein Data Bank, SwissProt, and Protein Identification Resource databases). The analysis identified 2,682 likely genes (including 52 RNA genes) in the *Halobacterium* NRC-1 genome, of which 1,658 coded proteins with significant matches to the databases. Of the matches, 591 were to conserved hypothetical proteins, and 1,067 were to proteins with known or predicted function.

For DNA repair, *Halobacterium* NRC-1 possesses two of the three genes involved in the guanine oxidization pathway, mutT and mutY. In addition, both the nucleotide and base excision pathways appear to be complete as copies of the uvrABC nuclease and uvrD helicase, and endonucleases and glycosylase genes are present. Two of the three genes of methyl-directed mismatch repair were found, mutL and mutS (three copies), but the nuclease gene mutH was missing. The *E. coli*-type dam methylase (recognizing GATC) is absent in NRC-1. However, a putative CTAG-specific methylase gene is present, which has also been found in *Methanobacterium thermoformicicum*.

Repair genes similar to those in yeast are present in *Halobacterium* NRC-1, including rad2, rad3, rad24, and rad25. Several of these proteins appear to be active in the excision repair pathway. Products of rad3 and rad25 have been identified as repair helicases and Rad2 is a single-stranded DNA endonuclease. This suggests that *Halobacterium* NRC-1 has developed multiple pathways to repair UV-induced damage as a means for survival. Cell-cycle genes in *Halobacterium* NRC-1 include five copies of cdc48, one of which is on pNRC200. The search for genes encoding proteins involved in recombination yielded two RadA genes, with homology to both the yeast protein Rad51 and the *E. coli* protein RecA.

The *Halobacterium* NRC-1-predicted proteome was compared with 11 other complete microbial genomes by using the DARWIN suite of programs. The results shown closest similarities to *Archaeoglobus fulgidus* and *Methanococcus jannaschii*. The NRC-1-predicted proteins were also similar to the Gram-positive bacterium, *B. subtilis*, more than to any other bacteria, and displayed a large number of unique homologs with the radiation-resistant bacterium *Deinococcus radiodurans*, suggesting that NRC-1 may have acquired a substantial number of genes from certain bacteria, possibly by lateral gene transfer. A more detailed comparative genomics investigation should provide further insights into evolutionary and adaptative forces operating in this extremophile. Because this halophile is amenable to experimental analysis by using a battery of approaches such as gene knockouts, DNA arrays, and proteomics, future studies should yield significant insights into the functions of conserved unknown and hypothetical genes among the archaea. Moreover, because the halophilic proteins are highly negatively charged with enhanced solubility, they lend themselves readily to the determination of high-throughput three-dimensional structure by experimental and theoretical approaches (structural genomics). Also, this system should serve as an excellent model of aspects of eukaryotic biology, e.g., DNA replication, transcription, and translation and DNA repair. Comparison of a halophile genome to other prokaryotic genomes should lead to a better understanding of microbial adaptation to extreme conditions, such as hypersalinity, damaging radiation, and an oxidizing atmosphere.

11. DNA Repair Transcriptomics in *Saccharomyces cerevisiae* Yeast

Many of the genes involved in various aspects of DNA repair were first identified in yeast. Completion of the yeast genome now presents the possibility of discovering further genes that underpin DNA repair, through, among others, genome-wide expression analysis in response to DNA-damaging agents. Damaging cells by physical and chemical carcinogens elicits significant changes in transcript level for more than 2,500 of *S. cerevisiae's* ~6,200 genes. There appear to be groups of genes that are specifically responsive to consecutive damaging agent and these may turn out to represent unique signatures for each agent. By far the largest category of responsive genes is genes of unknown function, and the next largest categories include those for protein and mRNA metabolism.

Ooi et al., in their screening of yeast genome for new DNA-repair genes, used a special collection of yeast deletion mutants [7]. Most of 6000 yeast open reading frames (ORFs) have been systemically disrupted, replaced by KanMX cassette (which results in resistance to the antibiotic) and marked by a special barcode consisting of two 20-nucleotide sequences flanking KanMX gene. The authors used a plasmid repair assay, in which double strand breaks can be generated within a region of a plasmid that is not homologous to chromosomal sequences. Therefore homologous recombination could not be used to repair DNA damage and non-homologous end-joining (NHEJ) mechanism was preferred. In this way they identified most of the known NHEJ genes, providing a clear "proof of principle" for this method. Moreover, they were able to identify new genes probably important for NHEJ pathway although it is not immediately evident how they might function. One ORF NEJ1, which codes for protein previously shown to interact with Lig4 complex was discovered in this way. Ooi et al show in their paper that deletion of this ORF results in a defect in NHEJ. This "post-genomic" screening strategy has clearly demonstrated its utility in identifying genes that it would have been difficult to discover with conventional genetic screens.

Whole-genome technologies in model organisms such as the yeast *Saccharomyces cerevisiae* offer a unique opportunity to study the global cellular response to inhibitors of this highly conserved pathway and to probe the specificity and possible secondary effects of these inhibitors. Fleming et al., combined two complementary approaches, whole-genome transcript profiling and a highly multiplexed competitive growth assay (population genomics), to characterize the cellular response to proteasome inhibition [2]. It was the first application for drug characterization of the bar-coded homozygous diploid strains coupled with chip technology. Clustering analysis identified four main groupings of up-regulated genes. One of the clusters consists of 215 genes whose induction depends on one protein - Rpn4p. This cluster contains essentially all of the proteasome subunit genes as well as many genes involved in ubiquitination, establishing that Rpn4p functions to mediate the induction of ubiquitin–proteasome genes in the face of proteasome inhibition. Strikingly, nearly all of these genes fall into only six functional classes on the basis of their cellular activities. It is intriguing that transcript profiling has shown that the nucleotide excision repair pathway is under Rpn4p control, whereas the mutants identified by population genomics function in recombination repair. Furthermore, there are links between protein ubiquitination and the postreplicative DNA repair pathway. The authors drew the conclusion that probably, in the face of proteasome inhibition, the burden of DNA repair is shifted to the recombination repair pathway that is not known to be ubiquitin-dependent. Alternatively, proteasome inhibition may induce genomic instability by perturbing mitotic events, and the mutants in this group may be hypersensitive to this effect.

12. DNA Repair Proteomics in *Saccharomyces cerevisiae* Yeast

With the advent of ultrasensitive mass spectrometric protein identification methods, it is feasible to identify directly protein complexes on a proteome-wide scale. Ho et al., using the budding yeast *Saccharomyces cerevisiae* as a test case, presented an example of this approach, which they term high-throughput mass spectrometric protein complex identification (HMS-PCI) [3]. A similar approach, where protein complexes "take a bait", was used by Gavin et al., in their analysis of the yeast proteome [1].

DNA damage response (DDR) includes DNA repair processes and checkpoint pathways that dictate cell cycle progression, transcription, protein degradation and DNA repair itself. The global DDR network revealed by HMS-PCI contained many known interactions as well as many new interactions of probable biological significance. Ho et al. constructed an initial set of 725 yeast bait proteins, from which they identified 3,617 interactions involving 1,578 different proteins. Among others they used 86 bait proteins that are implicated in the DNA-damage response, allowing them to delineate much of the yeast damage-response network. In particular, they reveal many regulators and targets of the protein kinase Dun1, and a possible role for the DNA-repair protein Rad7 in processes of targeted protein degradation. Most of the interactions identified did not depend on treatment with exogenous DNA-damaging agents, perhaps reflecting the fact that low level DNA damage normally occurs during replication. Examples of known interactions include: the replication factor C complex (RFC, Rfc1–5) and the RFCRad24 subcomplex, as well as the PCNA-like (PCNAL) Mec3–Rad17–Ddc1 complex, both of which transduce DNA damage signals; part of the Mms2–Ubc13–Rad18 post-replicative repair (PRR) complex; and the Mre11–Rad50–Xrs2 (MRX) complex that mediates double-strand-break repair by homologous and non-homologous mechanisms. The authors also recovered nearly all known nucleotide excision repair (NER) factors in their dedicated subcomplexes. The Rad4–Rad23 interaction (NEF2) was not found, but they nevertheless detected an association between Rad4 and NEF1, a known interaction among NER factors. There are many more clues in these results worth of further studies through comparative proteomics.

The approach taken by Gavin et al., and Ho et al., is clearly powerful, but it does have drawbacks. Both groups find a significant number of false-positive interactions, while failing to identify many known associations. Gavin et al., estimate that 30% of the interactions they detect may be spurious, as inferred from duplicate analyses of 13 purified complexes. Ho et al., meanwhile, did not detect nucleotide excision repair factor-2, a tight complex that contains the well- characterized DNA-repair proteins Rad4 and Rad23. So, as in most large-scale studies, these results are imperfect. It will be essential to integrate data from many different sources to obtain an accurate understanding of protein networks.

Proteomic studies such as these have generated a huge volume of exciting data. Yet — setting aside the problem of false positives and negatives — there is much still to be learned before we have a comprehensive knowledge of functional pathways within even a model organism such as yeast. Although feasible, the characterization of all remaining interactions will almost certainly be labor intensive [4].

References

[1] Gavin AC et al. Functional organization of the yeast proteome by systematic analysis of protein complexes. Nature 415, 141 - 147 (2002)

[2] Fleming et al. Complementary whole-genome technologies reveal the cellular response to proteasome inhibition by PS-341 Proc. Natl. Acad. Sci. USA. 2002 February 5; 99 (3): 1461-1466

[3] Ho Y et al. Systematic identification of protein complexes in Saccharomyces cerevisiae by mass spectrometry. Nature 415, 180 - 183 (2002)

[4] Kumar A and Snyder, M. Proteomics: Protein complexes take the bait. Nature 415, 123 - 124 (2002)

[5] Lodish, H et al. Molecular Cell Biology. 4th ed. New York: W H Freeman & Co 2000

[6] Makarova KS et al. Genome of the extremely radiation-resistant bacterium *Deinococcus radiodurans* viewed from the perspective of comparative genomics. Microbiology and Molecular Biology Reviews Vol. 65, No. 1, 44-79 (2001)

[7] Ooi SL et al. A DNA microarray-based genetic screen for nonhomologous end-joining mutants in *Saccharomyces cerevisiae*

[8] Wailap Victor Ng et al. Genome sequence of Halobacterium species NRC-1 Proc. Natl. Acad. Sci. October 24; 97 (USA. 2000 22): 12176–12181

Toxicogenomics and Proteomics
J.J. Valdes and J.W. Sekowski (Eds.)
IOS Press, 2004

Non-Ribosomal Peptide Synthetases for Production of Bioactive Peptides with Syringomycin Synthetase as an Example

N. Leyla ACAN
Hacettepe University, Faculty of Medicine, Department of Biochemistry, 06100 Ankara, Turkey

Abstract. Microbial toxins with peptide and protein structure have a potential to be used as pharmaceuticals such as antibiotics, immunosuppressive, chemotherapeutics, pesticide, etc. Most of these peptides are synthesized by non-ribosomal peptide synthetases. These synthetases are large molecular weight multienzymes that recognize and activate amino acids and synthesize peptide bonds following a so called "thiotemplate mechanism". This manuscript summarizes the current knowledge on the non-ribosomal peptide synthetases that produce a diversity of natural products, one of which is syringomycin. *Pseudomonas syringae* pv. *syringae* produce several secondary metabolites including syringomycins, and syringopeptins. These bacterial phytotoxins are cyclic lipodepsipeptides which are recently shown to be synthesized by non-ribosomal peptide synthesis mechanism. Syringomycins are shown to have immunosuppressive activity in addition to their phytopathological and antifungal properties. We have shown that they also have antimycobacterial properties. Syringomycin synthetase was purified 21.7 fold from Pseudomonas *syringae* pv. *syringae* B359 extracts with a yield of 8.7 %. At the final step two protein bands with molecular weights of 1200 and 740 kDa were obtained with SDS-PAGE. The purified protein had several properties indicating that it is a non-ribosomal peptide synthetase.

1. Introduction

Secondary metabolites are the products of specialized pathways that operate under certain conditions of stress, such as starvation. Unlike the products of intermediary metabolism, these substances are not essential for growth and reproduction. However they have several important functions for the organism. Some of the functions of the secondary metabolites may be to serve as a means of exportation or detoxification of waste products; or to participate in the defence or communication of the organism [1]. Apart from the specialized function for the organism they are produced by, the metabolites sometimes have useful pharmaceutical functions as antibiotics, chemotherapeutics, pesticides, immunosuppressive, antilipolytic agents, etc. [2]. Many of the microorganisms isolated from soil have antibiotic activity, actinomycetes being the most prolific producers among them. Antibiotics usually exert their effects by interfering with the metabolic pathways. Most of the time, however, this interference with the universal metabolic pathway causes a toxicity for the organism, preventing the therapeutic use. Animal toxicity may also be a result of a reaction which is different from the microbial reaction that the antibiotic specifically inhibits. Some antibiotics have a potential use in chemotherapy of cancer, since they have growth inhibitory activity in eukaryotes.

As with the diversity of their functions, secondary metabolites have diverse chemical structures, hence diverse pathways for their synthesis. The subject of this article is the summary of the knowledge of the non-ribosomal syntheses of the secondary metabolites with oligopeptide structure (non-ribosomal peptides) with a special example of syringomycins; and to summarize this laboratory's experience with syringomycin and syringomycin synthesis.

2. Non-ribosomal Peptides

Certain bioactive polypeptides are produced by the ribosomal protein synthesis system. These systems use the usual ribosomal mechanism and the peptides contain standard amino acids observed in all proteins. On the other hand, a non-ribosomal system which was proposed by F. Lipmann in 1971 [3], uses a mechanism which resembles fatty acid synthesis. In this system, peptide compounds are synthesized by high molecular weight multienzymes (peptide synthetases) following a so-called "thiotemplate mechanism".

The first nonribosomally synthesized peptides discovered were gramicidin S [4] and tyrocidin [5]. In the following years many other peptides with diverse pharmacological activity were found to be produced by the same mechanism. Some of the other examples of bioactive non-ribosomal peptides are actinomycin from *Streptomyces clavuligerus* [6], enniatin from *Fusarium oxisporium* [7], cyclosporin from *Tolypocladium inflatum* [8], surfactin from *Bacillus subtilis* [9], and well known antihypercholesterolemic lovastatin from *Aspergillus terreus* [10]. There are several modifications in the structures of these peptides. For example actinomycin and surfactin have lactone rings, enniatin is a depsipeptide which contains hydroxy acids linked by ester bond to amino acids, and lovastatin is a mixed peptide-polyketide compound. The knowledge of non-ribosomal peptides is still expanding and new compounds with novel pharmaceutical activities are being discovered each day. In addition to the currently known function and industrial use of these metabolites, discovery of new functions is also possible. For example, cyclosporin originally was used as an antifungal agent but, when its potent immunosuppressive activity was shown, it became the most important chemical used following tissue transplantations [11]. Many other examples can be found in references [1,2,12-17].

Non-ribosomal Peptide Synthesis

In this process, the constituent amino acids of the peptide are activated by synthetases as thioesters by consuming ATP. Peptide bond formation and chain elongation proceeds with the aid of a covalently bound phosphopanteteine residue, similar to fatty acid synthetase. In other words, peptide synthetases recognize and activate amino acids and synthesize peptide bonds. In addition to the usual L-amino acids found in proteins, D-amino acids or other unusual amino acids, or fatty acids are also used as building blocks in this type of synthesis. In recent years several excellent reviews on the mechanism of action of peptide synthetases have been published, some of which are listed in the references. The following section will be a brief summary of these reviews [12-17].

3. Structure and Functioning of Peptide Synthetases

Peptide synthetases have several features in common. They are made up of subunits, which can be called modules. Each module, approximately 1000 amino acids long, and 120 kDa in molecular mass, is specific for the addition of a particular amino acid. The number of modules on the peptide synthetase is equal to the number of building blocks of the peptide formed. The genes coding for the peptide synthetases are organized in large gene clusters of 20 kb or more, and the size coding for a single module is about 3 kb. The order of the modules usually indicate the sequence of the peptide synthesized. In this way the modules serve as the templates for the peptide synthesis. The simplest modules have an adenylation domain, a peptidyl carrier domain, and a condensation domain. A multiple carrier model is the currently accepted model for peptide bond formation by peptide synthetases.

Adenylation Domain

The adenylation domain is approximately 500 amino acids long. Since it recognizes and activates amino acids, its function may be compared to the activation of amino acids to make aminoacyl transfer RNAs in the ribosomal system. This domain has several highly conserved regions and an amino (imino or hydroxy) acid recognizing region. When the amino acid is recognized by the specific region of the domain, it is activated by an adenylation reaction at the expense of ATP. The reaction that take place at the adenylation domain of the module can be shown as follows, where aa is amino (imino or hydroxy) acid.

Adenylation reaction:

$$E_{(adenylation)} + aa + ATP + Mg^{2+} \Leftrightarrow E_{(adenylation)}(aa - AMP) + Mg^{2+} + PP_i \quad (1)$$

Peptidyl Carrier Protein Domain

In the next step, adenylated amino acid is covalently linked to the terminal sulphydryl of phosphopantetheine group at the peptidyl carrier protein domain (or thiolation domain), by thioester formation reaction, leaving AMP.

Thioester formation reaction:

$$E_{(adenyl.)}(aa - AMP) + E_{(carrier)}\text{-SH} \Leftrightarrow E_{(carrier)}\text{-S-aa} + E_{(adenyl.)} + AMP \quad (2)$$

Peptidyl carrier protein domain is about 80 amino acid long and its function is similar to the acyl carrier protein of fatty acid synthetases and polyketide synthetases.
This domain contains a 4'-phosphopantetheine group covalently attached to the hydroxyl group of a seryl residue, which functions as a swinging arm and carries amino acids to condensation domains.

Condensation Domain

Peptide bond formation reaction is catalyzed by the condensation domain, which is about 450 amino acids long. With the help of the phosphopantetheine group, thioester bound amino acid is carried to the acceptor site of the upstream condensation domain where it condenses with the amino acid or peptidyl group on the donor site, which was in turn carried there by the phosphopantheteine carrier of the previous module.

Condensation (chain initiation) reaction:

$$E_{(carrier\ 1)}\text{-S-aa}_1 + E_{(carrier\ 2)}\text{-S-aa}_2 \Rightarrow E_{(carrier\ 1)}\text{-SH} + E_{(carrier\ 2)}\text{-S-aa}_2\text{-aa}_1 \quad (3)$$

The newly formed peptide thioester is moved to the donor site of the downstream condensation domain to be linked with the next amino acid. Peptide bond formation and chain elongation continues by the stepwise addition of amino acids to the growing chain.

Chain elongation reaction:

$$E_{(carrier\ n-1)}\text{-S-aa}_{n-1}\text{-aa}_{n-2}\text{---aa}_1 + E_{(carrier\ n)}\text{-S-aa}_n \Rightarrow$$
$$E_{(carrier\ n-1)}\text{-SH} + E_{(carrier\ n)}\text{-S-aa}_n\text{-aa}_{n-1}\text{-----aa}_1 \quad (4)$$

The modules usually have additional regions such as (L to D) epimerization or N-methylation to modify the amino acids. By this way several D- or modified amino acids participate in the peptides formed by this mechanism.

Thioesterase Domain

Termination of peptide synthesis and the release of the completed peptide from the synthetase is achieved by the thioesterase domain, which locates at the end of the assembly line. This domain contains about 250 amino acids.

Chain termination reaction:

$$E_{(carrier\ n)}\text{-S-}aa_n\text{-}aa_{n-1}\text{-----}aa_1 + OH^- \Rightarrow E_{(carrier\ n)}\text{–SH} + aa_n\text{-}aa_{n-1}\text{-----}aa_1 \qquad (5)$$

4. Toxins Produced by Pseudomonas syringae pv. Syringae

Pseudomonas syringae pv. syringae produces several phytotoxins such as syringomycins, syringotoxins, syringopeptins and syringostatins [18]. All of these compounds are cyclic lipodepsipeptides. B359 strain, which we used in our research is known to produce only syringomycins and syringopeptins. Syringomycins (SR) (Figure 1) are cyclic lipopeptides that contain a hydroxy fatty acid and a nonapeptide moiety with several unusual amino acids such as 4-chlorothreonine, D-serine, 2,4-diaminobutyric acid and 2,3-dehydrobutyric acid. N-terminal serine is hydroxylated by a hydroxy fatty acid and O-acetylated by the C terminal carboxyl of 4-chlorothreonine. Main types of syringomycins such as SRa, SRe and SRG differ by their fatty acid component. The fatty acid components of these SR's are 3-hydroxy decanoic acid, 3-hydroxy dodecanoic acid and 3-hydroxy tetradecanoic acid, respectively [19].

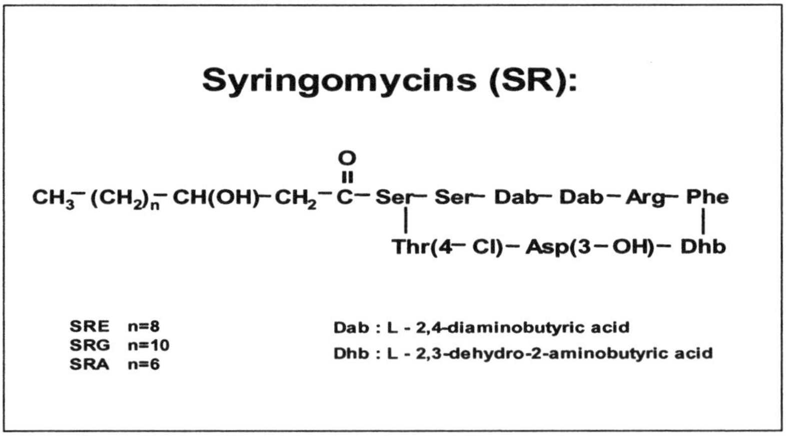

Figure 1. Structure of Syringomycins (Modified from reference 19).

Syringopeptins contain 22 or 25 amino acid residues including some unusual amino acids such as diaminobutyric and dehydrobutyric acids in the oligopeptide moiety. They have either A or B forms depending on the hydroxyfatty acid they contain. In A forms fatty acid is 3-hydroxy decanoic acid and in B forms, it is 3-hydroxy dodecanoic acid (20).

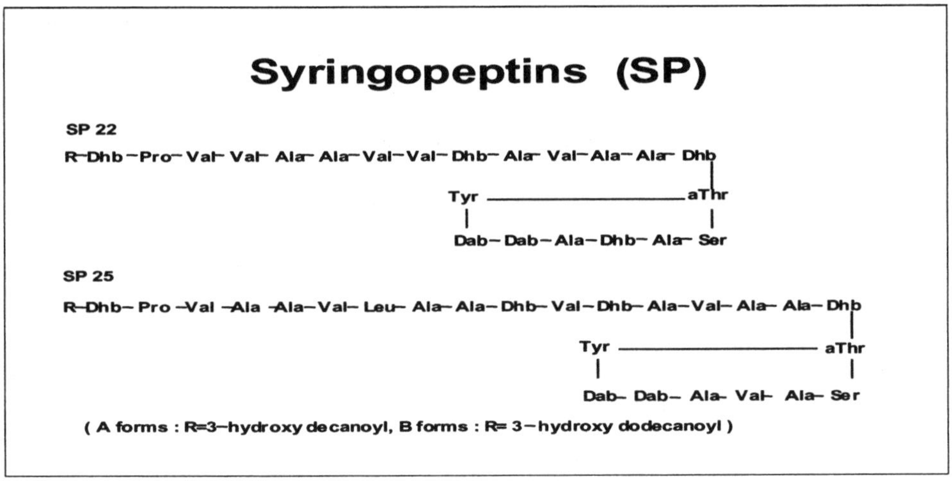

Figure 2. Structure of Syringopeptins (Modified from reference 20).

Both SR and SP are phytotoxic compounds. Both are pore forming cytotoxins that cause necrosis in plants. Owing to their amphipatic structures, they also have surface active properties [21].

Pharmaceutical Properties of Syringomycins

In addition to the phytotoxic properties, syringomycins -especially SRE- has several pharmaceutical properties. Recent increases in fungal infections and increasing fungal resistance to the antifungal chemicals stimulated the search for the development of new antifungals.

Sorensen et al. had evaluated the in vitro antifungal activity of SRE obtained from B301 strain against several medically important isolates and showed that SRE has strong activity against yeast with a MIC value as low as 2.5 μg/ml [22]. We also obtained comparable activity against *Candida krusei* and *Candida parapsilosis* with SRE produced by B359 strain [23]. Another important biologic activity of SRE is its immunosuppressive activity on proliferation of human blood lymphocytes. It was shown that SRE by itself had no effect, but mitogen-induced lymphocyte proliferation was significantly suppressed [24]. SRE has a potential to be used as an immunosuppressive compound.

In addition to these biological activities, we observed that the lipodepsipeptides produced by *Pseudomonas syringae* pv. *syringae* have strong antimycobacterial activity towards *Mycobacterium tuberculosis.* In order to investigate the antimycobacterial activities of individual lipodepsipeptides, acetone extract of these toxins were separated by cation exchange chromatography on Dowex 50W and the fractions were investigated for antimycobacterial activity against *Mycobacterium smegmatis.* MIC values found were between 1.5-3.2 μg/ml, which is comparable to some primary drugs for tuberculosis.

Among the lipodepsipeptides, Syringomycin E appears to be the most potent antimycobacterial agent [25]. SRE is cytotoxic to mammalians, causing lysis to sheep erythrocytes [22] and killing HeLa cells [26], although at higher concentrations as compared to MIC values obtained for their antifungal activities.

Just as the increase in fungal infections and increasing fungal resistance to the antifungal chemicals, tuberculosis insidence is also increasing and emerging resistance to classical drugs makes it a major health problem throughout the world. One third of World's population is infected with *Mycobacterium tuberculosis* and more than three million people die each year because of it. New drugs are urgently needed for effective treatment [27]. For this reason, we think that antimycobacterial activity of the lipodepsipeptides of *Pseudomonas syringae* is a potentially important finding. Once effective compounds are detected, they can be engineered to obtain more efficient and less toxic substances.

Purification and Characterization of Syringomycin Synthetase

Recently it was shown that syringomycins are synthesized by non-ribosomal thiol-template mechanism [28, 29]. Working with B301 strain of these bacteria and using a molecular biology approach, these compounds are recently proved to be synthesized by non-ribosomal peptide synthesis mechanism [30]. It was shown that, the 37 kb *syr* gene cluster of *Pseudomonas syringae* pv. *syringae* encodes four proteins (SyrB1, SyrB2, SyrC and SyrE) involved in biosynthesis. Gene products of Syr D and syrP are thought to function in secretion and regulation, respectively. 28.5 kb syrE is involved in the eight modules (from syrE1 to syrE6) for eight amino acids of SR and SyrB is for the ninth amino acid, threonine. We followed a biochemistry approach to purify the syringomycin synthesizing enzyme (syringomycin synthetase) from *Pseudomonas syringae* pv. *syringae* B359 extracts. Bacteria were grown in minimal medium and SR and SP production phase were determined. After the secretion conditions were optimized, incorporation of radioactively labeled amino acids into these substances were shown [31].

The purification method included ammonium sulfate fractionation (55%), gel filtration on Ultrogel AcA 34 (2x48 cm), butyl agarose (2x7.5 cm) and Phe-sepharose (1x3 cm) affinity chromatography steps. At the end of the purification steps, the enzyme was purified 21.7 fold with a yield of 8.7 %. Specific activity of the purified enzyme was 0.352 U/mg [32]. Due to the very high molecular mass, the enzyme was eluted with the void volume of the gel filtration column.

At the end of the hydrophobic chromatography step on butyl agarose chromatography, there was a possibility to copurify SR and SP synthesizing systems since SR and SP are structurally very similar compounds. It was observed that the butyl agarose column eluates catalyze the activation of the constituent amino acids of both syringomycin and syringopeptin via adenylation reaction with ^{32}P labeled PP_i (reverse of the reaction 1) and thioester formation reaction with ^{14}C labeled amino acids (reaction 2). In order to separate these closely related lipodepsipeptides synthesizing systems from each other and selectively purify SR synthetase, we made use of the presence of phenylalanine only in SR. Assuming that only SR synthetase has a recognition site for phenylalanine, a covalently bound phenylalanine containing affinity column was prepared. SDS-PAGE of the eluates of this column showed two protein bands with molecular weights of 1200 and 740 kDa (Figure 3). The molecular mass of the larger protein is in accordance with the molecular mass calculated from the base pairs of syrE which was about 1039 kDa. The smaller peptide is thought to be the degradation product of the native enzyme [30].

The final protein bands were able to use the constituent amino acids of only syringomycin in adenylation and thioester formation reactions. They showed strong cross reactivity with actinomycin synthetase specific antibodies. Covalent binding of ^{14}C labeled constituent amino acids into this protein was shown by autoradiography. Initiation of the syringomycin synthesis was also shown in cell-free extracts. Formation of the initiation product was possible only in the presence of the CoA ester of the hydroxyl fatty acid. Initiation product was not observed when the enzyme was incubated with 3-hydroxylauric acid alone or with a combination of 3-hydroxylauric acid and free CoA [31]. This requirement for the CoA derivative of the fatty acid resembles the biosynthesis of surfactin, which is also a lipopeptide [33].

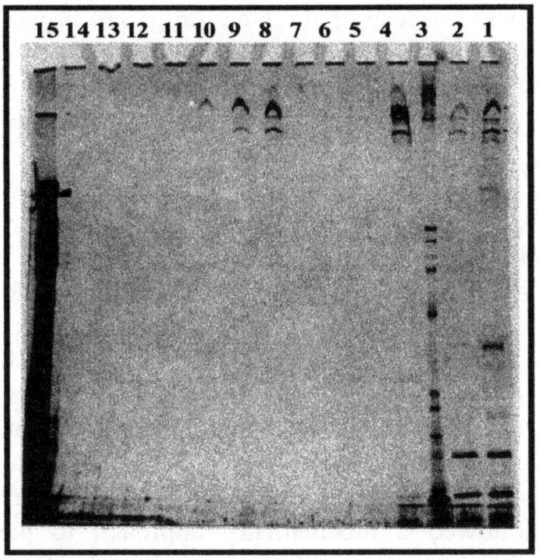

Figure 3.
SDS-PAGE after Phe-sepharose chromatography step. Lane 8-10 represent the final purified enzyme [32]

The properties of this purified protein strongly indicate that it is the non-ribosomal syringomycin synthesizing enzyme. Amino acid sequencing to show the conserved motifs of peptide synthetases remains to be elucidated.

5. Conclusion

Pharmaceutical and industrial importance of the non-ribosomally synthesized peptides makes this field rapidly expanding. Knowledge in the biosynthesis mechanism is rapidly increasing as well as the number of the novel peptides and the new properties of the already known ones are; syringomycin being just one example to these. Genetic engineering methods for these systems are also being developed. These developments eventually will lead to rational drug design of creating more potent peptides with fewer side effects.

Acknowledgements

The work summarized here was supported by the grants of NATO-B2, Deutsche Forschung Gemeinschaft and Scientific and Technical Research Council of Turkey. Purification of syringomycin synthetase was achieved at Max Vollmer Institute for Biophysical and Biological Chemistry at Technical University of Berlin, Germany in 1995-1996. I would like to express my thanks to PD Dr Rainer Zocher for allowing me to work in his laboratories and for his valuable collaborations. I also would like to extend my thanks to

Prof. H. Kleinkauf who was the Director of the Institute during my stay, and also to Prof. J. Salnikow, Drs U. Keller, H. von Doehren, J. Vater and all the other colleagues at the Institute for providing the excellent scientific and friendly atmosphere. Finally, I would like to thank my colleagues Dr. A. Stindl from Berlin and Ms E. Buber and Dr. T. Kocagoz from Ankara for their collaborations.

References

[1] L.C. Vining, Functions of Secondary Metabolites, *An nual Review of Microbiolog* 44 (1990) 395-427.

[2] R.L. Monaghan and J.S. Tkacz, Bioactive Microbial Products: Focus upon Mechanism of Action, *Annual Review of Microbiology* 44 (1990) 271-301

[3] F. Lipmann, Attempts to Map a Process Evolution of Peptide Biosynthesis, *Science* 173 (1971) 1435-1441

[4] W. Gevers *et al.* Peptidyl Transfers in gramicidin S biosynthesis from enzyme bound thioester intermediates, *Proceedings of the National Academy of Science USA* 63 (1969) 1335-1345

[5] R. Roskoski, Jr. *et al.* Isolation of enzyme bound peptide intermediates in tyrocidine biosynthesis, *Biochemistry* 9 (1970) 4846-4851

[6] A. Stindl and U. Keller, Tehe Initiation of Peptide Formation in the biosynthesis of actinomycin, *Journal of Biological Chemistry* 268 (1993) 10612-10620

[7] R. Zocher *et al.* Enniatin synthetase, a novel type of multi-functional enzyme catalysing depsipeptide synthesis in *Fusarium oxysporum, Biochemistry* 21 (1982) 43-48

[8] R. Zocher *et al.* Biosynthesis of Cyclosporin A, *Phytochemistry (Oxf.)* 23 (1984) 549-551

[9] B. Kluge *et al.* Studies on the biosynthesis of surfactin, a l,popeptide antibiotic from *Bacillus subtilis* ATCC 21332, *FEBS Letter* 231 (1988) 107-110

[10] L. Hendrickson, *et al.* Lovastatin biosynthesis in *Aspergillus terreus*: characterization of blocked mutants, enzyme activities and a multifunctional polyketide synthase gene, *Chemistry and Biology* 6 (1999) 429-439

[11] J.F. Borel and H.C. Gunn, Cyclosporin is a New Approach to Therapy Autoimmune Disease, *Annals of the New York Academy of Science* 475 (1986) 307-319

[12] H. Kleinkauf and H. von Doehren, A Nonribosomal System of Peptide Biosynthesis, *European Journal of Biochemistry* 236 (1996) 335-351

[13] R. Zocher and U. Keller, Thiol Template Peptide Synthesis Systems in Bacteria and Funghi, *Advances in Microbial Physiology* 38 (1997) 85-131

[14] H. von Doehren *et al.* Multifunctional Peptide Synthetases, *Chemistry Reviews* 97 (1997) 2675-2705

[15] H. von Doehren *et al.* The Nonribosomal Code, *Chemistry and Biology* 6 (1999) R273-R279

[16] M.C. Moffitt and B.A. Neilan, The Expansion of Mechanistic and Organismic Diversity Associated with Non-ribosomal Peptides, *FEMS Microbiology Letters* 191 (2000) 159-167

[17] T. Weber and M.A. Marahiel, Exploring the Domain Structure of Modular Nonribosomal Peptide Synthetases, *Structure* 9 (2001) R3-R9

[18] P. Lavermicocca et al. Biological Properties and Spectrum of Activity *Pseudomonas syringae* pv. *syringae* toxins, Physiol. Mol. Plant Pathol. 50 (1997) 129-140

[19] A. Segre *et al.* The Structure of Syringomycin A, E and G, *FEBS Letters* 255 (1989) 27-31

[20] A. Ballio *et al.* Syringopeptins, New Phytotoxic Lipodepsipeptides of *Pseudomonas syringae* pv. *syringae, FEBS Letters* 291 (1991) 109-112

[21] M.L. Hutchison and D.C. Gross, Lipopeptide Phytotoxins Produced by *Pseudomonas syringae pv. syringae*: Comparison of the Biosurfactant and Ion-channel Forming Activities of Syringopeptin and Syringomycin, *Molecular Plant-Microbe Interactions* 10 (1997) 347-354

[22] K.N. Sorensen et al. In vitro Antifungal and Fungicidal Activities and Erythrocyte Toxicities of Cyclic Lipodepsinonapeptides Produced by by *Pseudomonas syringae pv. syringae, Antimicrobial Agents and Chemotherapy* 40 (1996) 2710-2713

[23] M Ozalp et al. Partial Purification of Syringomycins and Syringopeptins, *Turkish Journal of Biochemistry* 25 (2000) 11-14

[24] V.K. Singh and J.Y. Takemoto, Suppression of Mitogen-Induced Lymphocyte Proliferation by Syringomycin-E, FEMS *Immunology and Medical Microbiology* 15 (1996) 177-179

[25] E. Buber *et al.* Antimycobacterial Activity of Lipodepsipeptides Produced by *Pseudomonas syringae pv. syringae, Natural Product Letters* 16 (2002) 419-427

[26] A.J. De Lucca *et al.* Fungal Lethality, Binding and Cytotoxicity of Syringomycin E, *Antimicrobial Agents and Chemotherapy* 43 (1999) 371-373

[27] A. Kochi, The Global Tuberculosis Situation and the New Control Strategy of the World Health Organization, *Tubercle* 72 (1991) 1-6

[28] L. Acan *et al.* Biosynthesis of Syringomycin, *Biological Chemistry Hoppe-Seyler* 376 (1995) S79
[29] J-H. Zhang *et al.* Analysis of *syrB* and *syrC* Genes of *Pseudomonas syringae* pv. *syringae* Indicates that Syringomycin is Dynthesiszed by a Thiotemplate Mechanism, *Journal of Bacteriology*177 (1995) 4009-4020
[30] E. Guenzi *et al.* Characterization of the Syringomycin Synthetase Gene Cluster, *Journal of Biological Chemistry* 273 (1998) 32857-32863
[31] N.L. Acan *et al.* Biosynthesis of Syringomycins and Syringopeptins, In: Abstract Book of the *XIII. National Biochemistry Congress Turkish Biochemical Society, Antalya, Turkey,* 26-30 March 1996, pp. B27-B28
[32] L. Acan *et al.* Purification of a Syringomycin Synthesizing Enzyme from of *Pseudomonas syringae* pv. *syringae* B359. In: Z. Demirbag (ed.) Proceedings of the 1st Eurasian Congress on Molecular Biotechnology, Karadeniz Technical University Press, Trabzon, 2001, pp. 10-14
[33] K. Ullrich *et al.* Cell-free biosynthesis of surfactin, a cyclic lipopeptide produced by *Bacillus subtilis, Biochemistry* 30 (1991) 6503-6508

Toxicogenomics and Proteomics
J.J. Valdes and J.W. Sekowski (Eds.)
IOS Press, 2004

Bacterial Genomics and Measures for Controlling the Threat from Biological Weapons

Jaroslav SPIZEK, Jiri JANATA, Jan KOPECKY and Lucie NAJMANOVA
Institute of Microbiology, Academy of Sciences of the Czech Republic
Videnska 1083, 142 20 Prague 4, Czech Republic
E-mail: *spizek@biomed.cas.cz*

Abstract. Bacterial genomics is used to detect rapidly pathogenic strains of microorganisms some of which could be used as biological weapons. Some examples are presented. Genomics of antibiotic producers and the knowledge of antibiotic gene clusters make it possible to design new antibiotics or their derivatives and new hybrid antibiotics. Genetic control of biosynthesis of lincosamide antibiotics in *Streptomyces lincolnensis* is described.

At the end of the 20th century microbiology appeared to lose its impetus and the main attention of the scientific community seemed to concentrate rather on molecular-biological study of eukaryotic organisms, particularly on the human genome. A certain lack of interest apparently also followed from the fact that infectious diseases seemed to be relatively easily curable by modern therapeutics, such as antibiotic. However, it is now known that infectious diseases caused by microorganisms are still the prime cause of mortality in the world despite the large number of antibacterial agents developed in the last century.

There are more than 30 new infectious diseases today that were unknown 20 years ago and some known infectious diseases re-appear with an increasing intensity. Widespread antibiotic use has resulted in a rapid spread of multi-drug resistant pathogens and it has become both a local and global health problem.

It appears that microbiology of the 21st century will have to deal with new and re-appearing infectious diseases. Apparently, antibiotic resistance is unavoidable and new anti-infectious drugs will be required. Microbiology can also offer interesting possibilities in the field of modern microbial biotechnologies including biodegradation and bio-remediation. Production of new biologically active compounds by genetically modified microorganisms rapidly develops and new drugs are in the pipeline. The development of microbiology in the 21st century can hardly be imagined without the simultaneous development of related fields such as immunology, combinatorial organic chemistry and biochemistry, analytical chemistry including most modern spectrometric methods, and bioinformatics, without which the evaluation of the enormous number of experimental data would be impossible.

Some of the new and reappearing infectious diseases and diseases caused by microorganisms that have not previously been considered pathogenic or by microorganisms resistant against antibiotics are listed in Table 1.

Numbers of antibiotic resistant bacteria permanently increase and many pathogens have become multidrug resistant (*e.g., Mycobacterium tuberculosis*). Some microorganisms that have previously been considered non-pathogenic cause serious nosocomial infections, particularly in intensive care units (*e.g. Acinetobacter baumanni, Pseudomonas aeruginosa, Bacteroides etc.*)

Table 1. Infectious microorganisms causing serious health problems

Bacteria	
Enterococcus faecium	*Pseudomonas aeruginosa*
Staphylococcus aureus	*Legionella pneumophila*
Acinetobacter baumanni	*Borrelia burgdorferi*
Streptococcus pneumoniae	*Helicobacter pylori*
Mycobacterium tuberculosis	*Vibrio cholerae*
Group A streptococci	*Haemophilus influenzae*
Escherichia coli O157-H7	
Viruses	
HIV	Ebola
Hantavirus	Mouth and foot virus
Prions	
BSA	
Eukaryotic microorganisms	
Plasmodium falciparum	Cryptococcal meningitis
Candida albicans	*Trichomonas*
Invasive pulmonary aspergillosis	*Cryptosporidium*
Pneumocystic carinii	

Of the microorganisms known today, some can be used as biological weapons. Biological weapons include living microorganisms and toxins (non-living poisons of biological origin) that are intended to be spread deliberately in aerosols, food or water to cause diseases, death or other harm to man, animals and plants.

They are potentially a serious threat. In the past, about 25 naturally occurring microorganisms (bacteria and viruses) and toxins have been considered for use as BW (Table 2).

Table 2. Natural biological agents

Bacteria	Viruses	Toxins
Bacillus anthracis	Venezuelan equine encephalitis	Botulinum
Yersinia pestis	Tick borne encephalitis	Ricin
Francisella tularensis	Spring/Summer encephalitis	*Clostridium perfringens*
Brucella species	West-Nile	*Staphylococcus* enterotoxin B
Vibrio cholerae	Ebola	Saxitoxin
Burkholderia pseudomallei	Marburg	Some mycotoxins
Burkholderia mallei	Smallpox	
Salmonella typhi	Influenza	
Rickettsia	Yellow fever	
Coxiella burnetti (Q fever)		
Rickettsia prowazekii		

Why does the concern over the possible use of biological weapons increase? There is a fear of the unknown effects of new agents that might be produced by genetic manipulation. Production of biological weapons is relatively cheap and easily concealed under the cover of peaceful activities, *e.g.*, in pharmaceutical companies.

The use of biological weapons is nothing entirely new. There are several examples of their application in the far and near past. Plague-infested corpses were used to break sieges in the middle ages. In 1932-1945 the Japanese deployed them against Chinese troops and civilians, apparently in an experimental fashion. Finally, anthrax spores have recently been used in the USA.

What are the countermeasures that can be used to prevent the possible uses of biological weapons? First of all, as far as the legislature is concerned, several international treaties have already been signed by a number of countries binding the involved countries not to use them. However, it is often difficult to control fulfillment of these treaties due to the absence of regular on-side inspections.

Therefore, new or modified ways of the control of both natural and intended infections have to be looked for. Sequencing of the human genome has recently been completed and significant progress has also been reached in sequencing of bacterial genomes. Over 100 bacterial genomes are being, or already have been, sequenced and assembled. This technological advance provided us with a wealth of data with which to compare microbial genomes, operons, genes, proteins and ultimately metabolic pathways. These comparative data already have been used in the study of basic processes in microbial cells, in the search for new targets in microbial cells, in the development of antibacterial drugs, for new or conserved surface antigens in the development of new vaccines and expression of new biosynthetic pathways.

One of the genomes of utmost importance is the *Bacillus anthracis* genome. Whole genome sequencing of a *Bacillus anthracis* Ames isolate ($pXO1^-$ $pXO2^-$) is complete, revealing a genome size of approximately 5.2 Mb. Many parts of the *B. anthracis* genome have a similar gene content and organization of the archetypal non-pathogenic *B. subtilis* and *B. halodurans* genome. At least 60 % of *B. anthracis* open reading frames have homologues of known *B. subtilis* genes. These include many spore-coat and spore-germination determinants and at least 23 components of a flagellum, although *B. anthracis* is considered to be non-motile. There are many genes without homologues in *B. subtilis* that might be important in anthrax infection such as several hemolysins and phospholipase genes. In the genome there are numerous copies of a conserved 16 bp part known as a target of the *Bacillus thuringiensis* positive regulator of extracellular virulence determinants plcR. However, it appears that the *B. anthracis plcR* gene contains deletions inactivating the gene. The pXO plasmids that contain the key virulence genes encoding toxin and capsule have recently been sequenced (Myers *et al.* 2002). Although the plasmids have probably undergone frequent rearrangements, there are few apparent cases of gene transfer between plasmid and chromosome suggesting a possible recent arrival of the episome into *B. anthracis*. The data should provide vital information mainly on evolution of pathogenicity determinants that will serve to design means against this dangerous pathogen. The authors from the Institute for Genomic Research also sequenced the Ames strain ($pXO1^+$ $pXO2^+$) isolated from a Florida bioterror victim. Two genome sequences revealed 60 new markers including single nucleotide polymorphisms (SNPs). A subset of these markers was tested on a collection of previously indistinguishable anthrax isolates to divide the strains into categories differing in virulence. The results demonstrate that the genome-based analysis of microbial pathogens will provide a powerful tool for the investigation of outbreaks of infectious diseases.

Francisella tularensis causing tularemia is another dangerous pathogen that can be used as a biological weapon. It is probably the most infectious pathogenic bacterium known, requiring as little as 10 organisms to cause disease. Most strains do not carry phages and plasmids. *Francisella novicida* F6168 seems to be the only member of the genus closely related to *Francisella tularensis* with a native plasmid, a 3990-bp pFNL10

(Pomerantsev *et al.* 2002). The entire plasmid was sequenced and six open reading frames were found that appear to be arranged into two operons. There are two distinct promoters similar to the *Escherichia coli* σ 70 promoter. Sequence analysis showed transcriptional terminators immediately downstream of the two operons. The 4442-bp pOM1 plasmid includes the pFNL10 replicon and tetracycline resistance gene regulated by the ORF5-ORF4 promoter. The fact that the tetracycline resistance gene is not expressed in pOM1-transformed avirulent *Francisella tularensis* mutants and is expressed in the virulent host suggests that pOM1 can serve as a tool for clarifying the virulence of this bacterium.

Strains of additional genera that could also serve as potential BWA microorganisms, such as the genera *Brucella*, *Salmonella etc.*, were also sequenced and sequencing of their genomes was completed.

In addition to genomic research, proteomics was applied in both *Francisella* and *Brucella* research. The proteome analysis was successfully used for the identification of different *Francisella* species, both non-virulent and virulent [3]. The proteomes of selected *Brucella* species were analyzed by utilizing current proteomic technology. Coupled with new and powerful system of data analysis, differentially expressed proteins between the non-virulent (vaccine) and virulent strains were categorized into several classes [4]. The proteomic research was based on the annotated *Brucella melitensis* genome [5].

The threat of attack on military and civilian targets with biological weapons is a growing international concern. Rapid detection and identification of biological weapons is essential for the treatment of outbreaks of infectious diseases. Both the genome and proteome analysis can rapidly yield valuable information on the type of infection, the nature of the infection and sometimes detection of the source of the infection.

Once they occur, the infectious diseases can be treated in several ways. Specific vaccines or antibiotics can be used, but really safe and efficient vaccines are often difficult to prepare and pathogenic bacteria readily develop antibiotic resistance. However, new antibiotics or at least new derivatives of known antibiotics can be prepared by using methods of molecular genetics and biology. Some approaches to be used for the treatment of outbreaks of infections diseases are listed in Table 3.

Table 3. Some approaches for the treatment of infectious diseases

1)	Search for new targets in pathogenic bacteria based on genome sequencing
2)	Production of new protective vaccines
3)	Use of probiotics to increase immunological response
4)	New hybrid antibiotics, mainly against resistant bacteria
5)	Metabolic engineering to improve producing strains of microorganisms

Antibiotics are secondary metabolites produced by various organisms. The term secondary metabolites was first used in plant physiology and has been used for compounds that do not play a distinct role in the vegetative development, and, hence, whose role is only secondary. In microorganisms, there usually exists an inverse relationship between the growth rate of the producing strain and secondary metabolite production. In addition, the secondary metabolite production is usually closely connected with morphological differentiation and only occurs when the growth limiting substrate in the cultivation medium has been utilized. The term secondary metabolite stands for compounds with most varied chemical structure and biological activity such as pigments, alkaloids, aromatic compounds, mycotoxins, antibiotics *etc.* In this paper the attention will mainly concentrate on antibiotics, *viz.* the genetic and physiological control of their biosynthesis.

The concept of antibiotic activity was first introduced as early as in 1889 by Paul Vuillemin, who used the term "antibiotic influences" to describe negative interactions among plant and animals. However, in the forties, Waksman coined the term “antibiotic” and described an antibiotic as a “chemical substance derived from microorganisms which has the capacity of inhibiting growth, and even destroying, other microorganisms in dilute solutions”. As we mentioned previously, natural antibiotics belong to a group of compounds called secondary metabolites, generally characterized by having structures that are unusual compared with those of intermediary metabolites, by being produced at low specific growth rates, and by the fact that they are not essential for growth of the producing organisms in pure culture. Antibiotics are thought, however, to be critical to the producing organisms in their natural environment, as they are needed both for survival and competitive advantage.

More than 10,000 natural antibiotics have been discovered and more than 100,000 partially and totally synthetic derivatives have been developed. The main producers of antibiotics are shown in Table 4.

Table 4. Main producers of antibiotic compounds

Producing organism	Estimated numbers of produced antibiotics
Actinomycetes	5 200
Other bacteria	1 100
Higher plants	3 200
Fungi	2 000
Animals	800
Algae	45
Mosses	10

Antibiotics can be classified according to their origin, biological activity, mode of action, chemical structure and biosynthetic pathway. Our knowledge concerning the genetic control of antibiotic biosynthesis is briefly summarized in Table 5.

Table 5. Genetic control of biosynthesis of secondary metabolites

1)	Genes coding for biosynthesis of a secondary metabolite are arranged in clusters on the chromosome
2)	They do not constitute a single operon but are organized in several transcription units
3)	Gene expression is controlled primarily at the level of transcription
4)	Genes coding for resistance to a given metabolite are closely linked with the cluster of biosynthetic genes or constitute its part
5)	Expression of biosynthetic genes and genes coding for resistance is controlled in a coordinated manner
6)	Specific regulatory genes are situated in the closest proximity of biosynthetic genes
7)	A central overruling regulatory circuit often exhibiting pleiotropic effects sometimes controls expression of genes coding for resistance (Spížek and Tichý 1995).

Some aspects of physiological control of secondary metabolite production are mentioned in Table 6. In fact, both the genetic and physiological controls are closely connected and it is often quite difficult to discriminate between them.

Table 6. Physiological control of biosynthesis of secondary metabolites

1)	Different types of feed-back inhibition
2)	Phosphate inhibition
3)	Carbon and energy source regulation
4)	Nitrogen source regulation
5)	Oxygen availability
6)	Autoregulatory compounds

There are numerous examples of feedback inhibition and many review articles on this topic have been published. Long ago, we were also able to demonstrate feedback inhibition in the biosynthesis of cycloheximide and related compounds and fungicidin in *Streptomyces noursei* [6].

Phosphate has long been known as an inhibitor of biosynthesis of secondary metabolites, at least at elevated concentrations. Apparently, phosphate plays an important role in the process. Although alot of work was devoted to this problem, the molecular mechanism by which many promoters of biosynthetic genes are regulated is still obscure. Either all promoters contain similar "phosphate" boxes or, alternately, a common DNA-binding protein must occur that in turn is regulated by phosphate. This protein may be a specific σ-factor or a protein interacting with RNA polymerase. Protein phosphorylation and protein kinases catalyzing it may play an important role here. As a matter of fact, the WD-repeat proteins of prokaryotic organisms first discovered in our laboratory and usually containing the WD-repeat domain, a DNA-binding domain and a protein kinase domain, may be involved in this type of control [7].

There are again numerous examples of carbon catabolite repression in the biosynthesis of secondary metabolites. It is the rule that the rapid growth supported by suitable concentrations of readily utilizable carbon and energy sources suppresses the production of secondary metabolites and that a slow utilization of carbon and energy sources and their good balance are favorable.

High concentrations of ammonia in the medium often suppress the biosynthesis of secondary metabolites and there is usually a clear-cut relationship with the primary metabolism, namely metabolism of branched-chain amino acids and biosynthesis of some secondary metabolites, *e.g.,* avermectins produced by *Streptomyces avermitilis*.

Aeration, an important physiological aspect of biosynthesis of secondary metabolites, has often been neglected but it clearly plays an important role. Long ago, it was demonstrated that the interruption of aeration in the early stages of *Streptomyces aureofaciens* cultivation resulted in a substantially decreased or fully suppressed tetracycline biosynthesis.

Many streptomycetes use small, lipid- and water-soluble γ-butyrolactones as hormones to trigger secondary metabolism and aerial mycelium formation. The best studied example, A-factor of *Streptomyces griseus*, is synthesized in a growth-dependent manner. Early in the cultivation, when the concentration of A-factor is low, an A-factor receptor (ArpA) binds and represses a still unknown gene(s) necessary for the onset of secondary metabolism and/or morphogenesis. As the culture becomes denser, the concentration of the A-factor raises to a critical level, at which it binds to ArpA. ArpA then dissociates from the DNA and thus switches on the transcription of the key gene *adpA* coding for a transcriptional activator. The cascade continues when the AdpA protein switches on gene *strA* encoding the pathway-specific transcriptional activator for streptomycin biosynthesis. StrA then activates transcription of all the streptomycin-biosynthetic genes.

In spite of the fact that many autoregulatory compounds were described and their mechanism of action was often investigated in detail, their role in general, particularly at the molecular level, is virtually unknown. They may act as messengers, enzyme stimulators, enzyme inhibitors *etc.*

Secondary metabolite product yields can be improved by optimizing nutritional and physical component of the fermentation medium, by genetic modification of the producing organism and, naturally, iterative combinations of the two processes.

A general method of undisputed value to strain improvement is chemical or physical mutagenesis followed by fermentation analysis to identify improved strains. Of the mutagens analyzed in comparative studies, N-methyl-N′-nitro-N-nitrosoguanidine (MNNG) was the most potent, followed by methyl methanesulfonate (MMS), 4-nitroquinoline-1-oxide (NQO), ethylmethanesulfonate (EMS), hydroxylamine (HA), and ultraviolet light (UV).

The potential applications of genetic recombination and efficient methods for genetic recombination by protoplast fusion have been available for some time, however, significant published examples of applications of recombination, at least in streptomycetes, are still lacking. Even if recombination is not as robust as random mutagenesis in strain development, it is useful for the construction of recombinants with specific useful traits that cannot be generated by mutagenesis. In this respect, it is useful to employ recombination as a complementary method to augment mutagenesis and selection. Genetic engineering is apparently the most promising approach for the strain improvement and biosynthesis of new antibiotic compounds.

Biosynthesis of many secondary metabolites is catalyzed by large multifunctional or multisubunit enzymes – the polyketide synthases. These enzymes can assemble small fatty acids such as acetic acid and propionic acid and their carboxylated derivatives into complex compounds. Biosynthesis of many relatively reduced polyketide antibiotics including erythromycin and avermectin is catalyzed by Type I, or modular PKSs consisting of several large multifunctional enzymes, each composed of modules that contain all the enzymatic functions required for processing of the individual fatty acid building blocks. There is a separate set of fatty acid synthase-related enzyme domains for each cycle of polyketide chain extension. For example, there are fourteen such "modules" of enzymes in the PKS catalyzing biosynthesis of rapamycin, a valuable immunosupressant. As the individual PKS modules contain a variable number of reductive enzymes and acyltransferases specific for propionate or acetate extender units, diverse products can arise from a single biosynthetic pathway. It has already been demonstrated that productive hybrid PKSs can be made by a surgery and replacement technique in which individual domains or sets of domains are substituted from one PKS or producing strain into another [8].

It is the aim of future work to understand the structure and mechanism of action of modular PKSs and on the basis of the knowledge obtained to design productive hybrids that can be used for the production of secondary metabolites of predictable structure.

As compared with modular polyketide synthases the Type II polyketide synthases are composed of several, usually monofunctional enzymes that catalyze the same action repeatedly. These are involved in biosynthesis of cyclic aromatic polyketides such as actinorhodin and daunorubicin.

If one of the enzymes of the PKS is deleted, or substituted by a similar enzyme from another organism, new secondary metabolites may be produced. This combinatorial biosynthesis approach, exploiting the interchangeable nature of many of the enzymes of cloned type II PKSs has already yielded a wide range of new compounds that can be tested for their possible therapeutic or industrial uses [9].

In our laboratory we adopted a simpler model for studying the biosynthesis of secondary metabolites, *viz.* biosynthesis of the lincosamide antibiotics.

Lincomycin consists of two parts, a sugar moiety (6-amino-6,8-dideoxy-1-thio-D-erythro-α–D-galactooctopyranoside - MTL) linked *via* a peptide bond to an amino acid moiety (trans-N-methyl-4-*n*-propyl-L-proline - PHA). A minor analog, lincomycin B (4′-depropyl-4′-ethyllincomycin), is also normally produced by some strains during fermentation. It contains ethylproline instead of propylproline. Clindamycin, a semi-synthetic 7-chloro derivative, is roughly ten times more active against Gram-positive bacteria, *e.g.*, clinical strains of staphylococci, streptococci and diplococci in particular, than lincomycin and is commonly used for therapeutic purposes (Fig. 1). As far as the mechanism of action is concerned, lincomycin blocks protein synthesis in sensitive strains by preventing binding of aminoacyl-tRNA to the mRNA-ribosome complex in the 50S subunit. Consistent with this mode of action, one mechanism of resistance to this drug has been shown to be target-site modification involving methylation of adenine residues in 23S RNA.

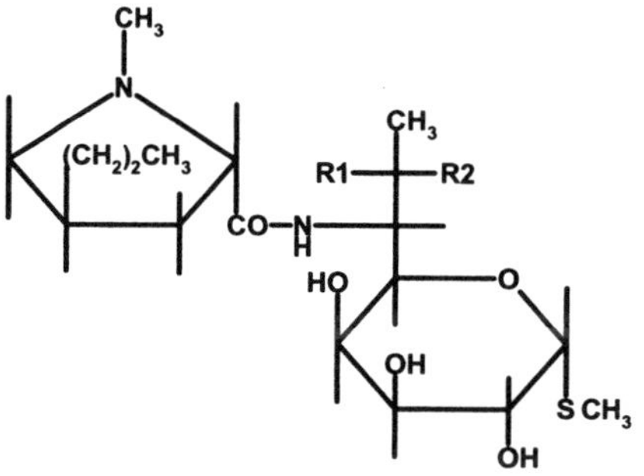

	R1	R2
Lincomycin	-OH	-H
Clindamycin	-H	-Cl

Fig. 1. Chemical structure of lincosamide antibiotics

The biosynthetic pathway leading to lincomycin is shown in Fig. 2. As mentioned previously, lincomycin A is composed of an amino acid subunit, PHA, and a sugar subunit, MTL. In summary, it was shown clearly that the primary metabolic precursor of PHA is L-tyrosine, with the N-methyl and terminal side-chain methyl groups added in reactions involving transfer from S-adenosylmethionine. The origin of MTL moiety is less certain, but it seems most likely to involve a nucleotide-activated hexose, with the amino group added *via* a transamination reaction and the S-methyl group added as a single unit. It can be seen from the scheme that the two basic precursors are linked together in an N-demethyllincomycin (NDL) synthetase reaction and that the ultimate step in the pathway is methylation, apparently catalyzed by NDL-transmethylase. In the amino acid branch of the pathway L-tyrosine is converted to L-dihydroxyphenylalanine (L-DOPA) that in a series intermediates, whose structures have so far been only partially clarified, yields PPL. We investigated mainly the onset of PPL biosynthesis and the final methyltransferase reaction.

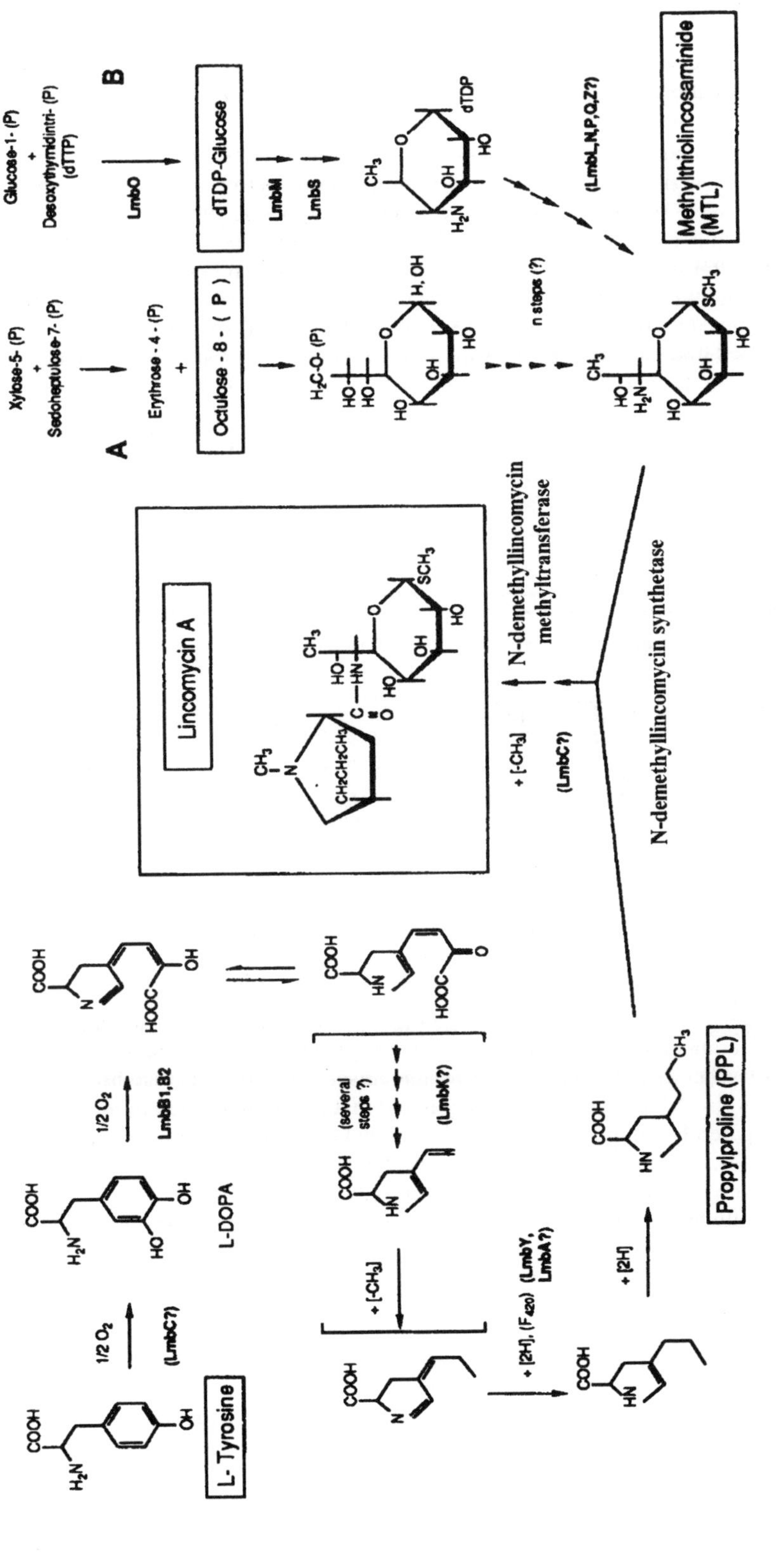

Fig. 2. Lincomycin biosynthetic pathway

Relatively recently, significant information has been obtained on the genetics and biochemistry of lincomycin biosynthesis. Most notable in this regard is the isolation, sequencing, and analysis, by Piepersberg and his coworkers, of a cluster of *Streptomyces lincolnensis* genes that probably encode the entire lincomycin biosynthetic pathway. We contributed to their results by identifying the genes coding for the onset of biosynthesis of the amino acid moiety and genes involved in the final steps of lincomycin biosynthesis.

The entire conserved region from *Streptomyces lincolnensis* was sequenced and found to contain 30 open reading frames (Fig. 3). Based on databases the functions encoded by several of them have been at least partially characterized and are clearly involved in lincomycin production. The figure is drawn according to the results of Peschke *et al.* [10] *Lmb* indicates biosynthetic genes, *lmr* stands for genes coding for functions that impart a lincomycin-resistance phenotype. However, a gene coding for a regulatory protein seemed to be missing. Such genes usually code for transcription activators. We looked for

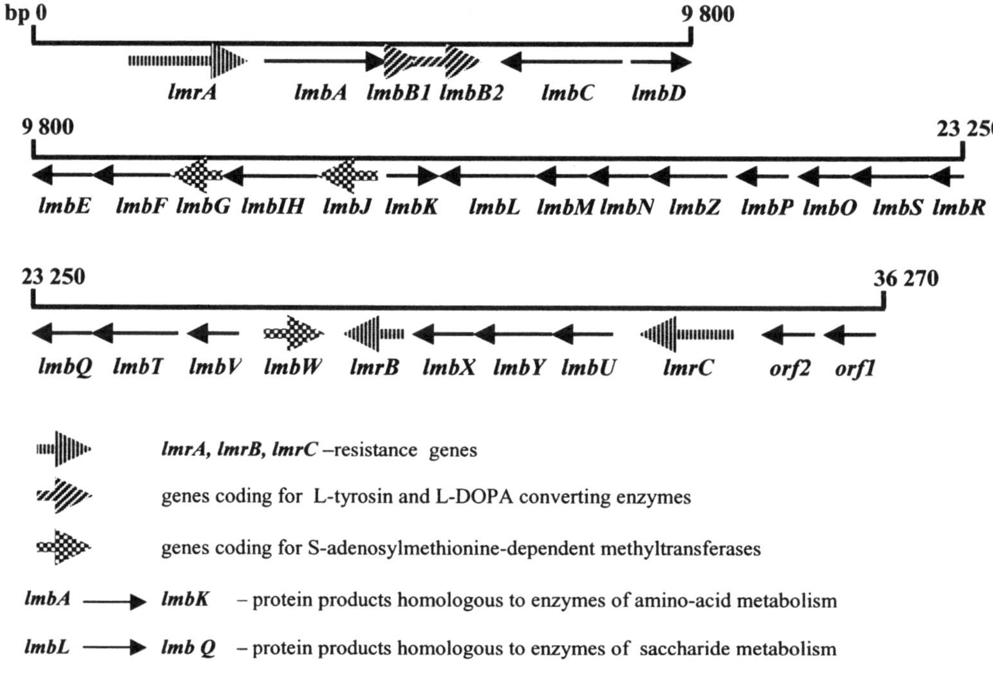

Fig. 3.

such a gene and found that the initial assumption of Peschke and coworkers [10] of separate *lmbH* and *lmbI* genes with so far unidentified function might have been wrong and that probably only a single gene *lmbIH*, coding perhaps for a regulatory protein, is involved [11].

The general scheme of gene clusters coding for biosynthesis of secondary metabolites need not necessarily always be true and, naturally, there are many exceptions to the rule. Biosynthesis of polyketides by different types of polyketide synthases may serve as examples. As a matter of fact, specific resistance genes are quite often missing in the gene cluster, although some of the genes coding for membrane transport proteins and ABC transporters may play such a role here.

When analyzing the lincomycin gene cluster sequence we detected a conserved motif of S-adenosylmethionine dependent methyltransferase also in the gene *lmbJ* product. We have chosen this gene, due mainly to its position in the gene cluster between the amino acid and sugar genes and similarity with some N-methyltransferases, as a likely ORF coding for the NDL-methyltransferase. When inspecting the chemical structure of lincomycin and celesticetin it can be seen that there are several methyl groups in both structures, however, there is only a single N-methyl group that is common for both compounds. Therefore, we hypothesized that genes coding for N-demethyllincomycin transmethylase should be present in genomes of both producers, *i.e.* in *S. lincolnensis* and *S. caelestis*. Oligonucleotide primers designed according to the known sequence of the *S. lincolnensis lmbJ* yielded, after amplification, a specific product with a correct size that was found both on chromosomal DNA of the lincomycin producer and on chromosomal DNA of *Streptomyces caelestis* producing a structurally similar celesticetin. The DNA probe with a sequence of the *S. lincolnensis lmbJ* also yielded a specific signal with DNA of *S. caelestis*. As the final N-methylation is the only common methyltransferase reaction in both related biosynthetic pathways, we assume that *lmbJ* and its *S. caelestis* homolog are involved in the final modification of the two lincosamide antibiotics. The *lmbJ* sequence of chromosomal DNA from *S. lincolnensis*, was amplified and detected in both *S. lincolnensis* and *S. caelestis* indicating that it can be involved in the conversion of N-demethyllincomycin to lincomycin. The PCR product was cloned, expressed in *E. coli* and purified. The enzyme activity of the LmbJ protein was then tested with chemically synthesized N-demethyllincomycin as substrate. It was demonstrated that the enzyme can really convert N-demethyllincomycin to lincomycin. The LmbJ protein was further characterized and it was shown that its pH optimum is 8.3, temperature optimum 31 ^{0}C, apparent molecular weight of the LmbJ monomer is 30 kDa, whereas calculated molecular weight of the native enzyme is 272 kDa The native LmbJ protein was further studied by gel filtration which yielded peaks of 70, 144 and 272 corresponding roughly to dimer, tetramer and perhaps octamer of the basic 30 kDa unit. Electron microscopy revealed four-fold symmetry complexes in negatively stained samples of the LmbJ protein.

What is the predictable future of genetic engineering of secondary metabolites? One can easily envisage that using genetic manipulation new producing strains of microorganisms will be designed synthesizing new secondary metabolites useful for potential or industrial uses. The available knowledge of the genome structure of *Streptomyces coelicolor* [12] indicates that genes coding for enzymes involved in the biosynthesis of different new metabolites are present that have never been previously detected in this bacterium. It appears that based on this knowledge new derivatives of antibiotics will be isolated. With the present knowledge of genome structure of some pathogenic bacteria new compounds will be looked for with the aim of specifically intervening with these new targets. Such compounds would then be used to design new drugs particularly against viruses and antibiotic-resistant bacteria. Numbers of antibiotic resistant microbial strains permanently increase and as Julian Davies pointed it out: "It is frightening to realize that one single base change in a gene encoding a bacterial β-lactamase can render useless US dol. 100 million worth of pharmaceutical research effort".

In spite of all the progress in rapid detection of microbial pathogens and significant research and development in the area of treatment of infectious diseases it is still the microbe that plays the most important role. At the end of this paper it is worth mentioning the Louis Pasteur statement: "In all situations the microbe will always have the last word." Whether we like it or not, it appears that the great chemist and microbiologist was right.

References

[1] G.S.A. Myers *et al.*, The *Bacillus anthracis* Ames genome sequence, Applications of Genomics and Proteomics for Analysis of Bacterial Biological Warfare Agents, NATO-ARW, Bratislava, Slovak Republic, July 24–28, 2002.

[2] A.P. Pomerantsev *et al.*, Genome analysis of *Francisella* specific plasmids probably leads to the host clarifying virulence, Applications of Genomics and Proteomics for Analysis of Bacterial Biological Warfare Agents, NATO-ARW, Bratislava, Slovak Republic, July 24–28, 2002.

[3] L. Hernychova *et al.*, Proteome of *Francisella tularensis*, Applications of Genomics and Proteomics for Analysis of Bacterial Biological Warfare Agents, NATO-ARW, Bratislava, Slovak Republic, July 24–28, 2002.

[4] C.V. Muyer *et al.*, Principles and applications of proteomics in *Brucella* research, Applications of Genomics and Proteomics for Analysis of Bacterial Biological Warfare Agents, NATO-ARW, Bratislava, Slovak Republic, July 24–28, 2002.

[5] V.G. DelVecchio *et al.*, The genome sequence of the facultative intracellular pathogen *Brucella melitensis*. *Proc. Natl. Acad. Sci., USA* 99 (2002) 443–448.

[6] J. Spizek and P. Tichy, Some aspects of overproduction of secondary metabolites, *Folia Microbiol.* **40** (1995) 43–50.

[7] L. Janda *et al.*, A deduced *Thermomonospora curvata* protein containing serine-threonine protein kinase and WD-repeat domains. *J. Bacteriol.* **178** (1996) 1487–1489.

[8] P.F. Leadlay et al., Engineering of complex polyketide biosynthesis – insights from sequencing of the monensin biosynthetic gene cluster, *J. Ind. Microbiol. Biotechnol.* **27** (2001) 360–367.

[9] C.R. Hutchinson and R. McDaniel, Combinatorial biosynthesis in microorganisms as a route to new antimicrobial, antitumor and neuroregenerative drugs, *Curr. Opin. Investig. Drugs* **2** (2001) 1681–1690.

[10] U. Peschke *et al.*, Molecular characterization of the lincomycin production gene cluster of *Streptomyces lincolnensis* 78-11. *Mol. Microbiol.* **16** (1995) 1137–1156.

[11] J. Janata *et al.*, Putative *lmbI* and *lmbH* genes form a single *lmbIH* ORF in *Streptomyces lincolnensis* type strain ATCC 25466, *Anthon. Leeuw.* **79** (2001) 277–284.

[12] M. Redenbach *et al.*, A set of ordered cosmids and a detailed genetic and physical map for the 8 Mb *Streptomyces coelicolor* A3(2) chromosome, *Mol. Microbiol.* **21** (1996) 77–96.

Toxicogenomics and Proteomics
J.J. Valdes and J.W. Sekowski (Eds.)
IOS Press, 2004

An Evaluation of Toxins and Bioregulators as Terrorism and Warfare Agents

Slavko BOKAN
MOD of the Republic of Croatia, Croatian Military Academy,
Laboratory for NBC Protection, HR-10000 Zagreb, Ilica 256 b, Croatia

Abstract. Bioregulators or modulators and toxins are biochemical compounds, such as peptides, that occur naturally in organisms. They are new class of weapons that can damage nervous system, alter moods, trigger psychological changes and kill. Within neuroscience over the last twenty years has been an explosion of knowledge about the receptor systems on nerve cells that are of critical importance in receiving the chemical transmitter substances released by other nerve cells. The potential military or terrorism use of bioregulators is similar to that of toxins. Together with increased research into toxins, the bioregulators have also been studied and synthesized. This paper presents evaluation of toxin warfare agents and bioregulators that can be used as terrorism delivery system or biological agents in hostile activities.

1. Introduction

Many biological agents have the capacity to cause disease and potentially be used to threaten civilian populations. The purpose of this paper is to provide information on biological toxins and bioregulators to military and health-care providers at all levels to help them make informed decisions on protecting from these agents. Bioregulators can act as neurotransmitters and modify neural response. Bioregulators are closely related to substances normally found in the body that regulates normal biological processes. Some examples of potential applications of bioregulators are to cause pain, as an anesthetic and to influence blood pressure.

These substances can also modified synthetically, whereupon they may obtain new properties. It is feasible to produce some of these compounds by chemical synthesis. It is apparent that the past decade has brought an enormous increase in knowledge about the pharmacology and structural biology of receptors.

In the last ten years considerable advances have taken place in this *in vitro* synthesis of peptides and already commercial production in large quantities of various pharmaceutical peptides are freely available. Synthetic derivatives or slightly modified forms of these compounds can have drastically altered toxic effects and these could be important in the development of new agents. Advances in discovery of novel bioregulators, especially bioregulators for incapacitating, understanding of their mode of operation and synthetic routes for manufacture have been very rapid. Some of these compounds may be potent enough to be many hundreds of times more effective than the traditional chemical warfare agents. Some very important characteristics of new bioregulators that would offer significant military advantages are novel sites of toxic action; rapid and specific effects; penetration of protective filters and equipment and militarily effective physical incapacitation. Peptide bioregulators are interesting regulatory molecules for many reasons. Their range of activity covers the entire living system, from mental processes (e.g.

endorphins) to many aspects of health such as control of mood, consciousness, temperature control, sleep, or emotions, exerting regulatory effects on the body.
Although the use of toxins and bioregulators in military conflicts has been a concern of military communities for many years, several recent events have increased the awareness regarding the potential use of these weapons by terrorists against civilian populations. Although no known illnesses resulted from these attempts to use these agents as weapons, the exploration of the use of these agents by this and other extremist organizations caused great concern regarding the potential vulnerabilities of civilians to such weapons.

2. Materials and Methods

Toxins are effective and specific poisons produced by living organisms. They usually consist of an amino acid chain, which can vary in molecular weight between a couple of hundred (peptides) and one hundred thousand (proteins). They may also be low-molecular organic compounds. Toxins are produced by numerous organisms, e.g., bacteria, fungi, algae and plants. Many of them are extremely poisonous, with a toxicity that is several orders of magnitude greater than the nerve agents.

Research literature suggests that we have discovered the majority of the "most toxic" (LD50 < 0.0025 mg/kg) naturally occurring toxins. Because they must be delivered as respirable aerosols, their toxicities and ease of production limit toxins' utility as effective MCBW.

Toxins are poisonous substances produced by the metabolic activities of certain living organisms, including bacteria, insects, plants, and reptiles. Toxins are still considered to be less suitable for dispersal on a large scale. Nonetheless, they could be used for sabotage or in specially designed inputs, e.g., against key persons. Since toxins have low volatility, they are dispersed as aerosols and then taken up primarily through inhalation. The new microencapsulation technology, which is easy to use, makes it possible to protect unstable toxins when dispersed. During recent years, discussions have started on the risk of bioregulators being used as CW agents. These types of substances do not belong to the group of toxins but are, nonetheless, grouped with them since their possible use is similar. They are closely related to substances normally found in the body and may be algogenic (causing pain), anaesthetic, or influencing blood pressure. A characteristic of them is that they are active in extremely low doses and frequently have rapid effects.

One example of this group of substances is Substance P, a polypeptide (molecular weight = 1,350 D) which is active in doses of less than one microgram. Substance P causes, for example, a rapid loss of blood pressure, which may cause unconsciousness. Fifteen to twenty of some 400 known toxins have the physical characteristics that make them threats against military forces as potential MCBWs. However, many toxins could be used in weapons to produce militarily significant/terrorist (psychological) effects-especially in poorly prepared civilian populations.

There are still many unknowns regarding toxins and their weaponization. Some toxins can be produced by molecular biologic techniques (protein toxins) or by chemical synthesis (low molecular weight toxins). A Mass Casualty Biological (toxin) Weapon (MCBW) is any toxin weapon capable of causing death or disease on a large scale, such that the military or civilian infrastructure of the state or organization being attacked is overwhelmed. A militarily significant (or terrorist) weapon is any weapon capable of affecting-directly or indirectly, physically or through psychological impact-the outcome of a military operation

From a public health standpoint, bioregulators which are less known, must be evaluated and prioritized in order to assure appropriate allocation of the limited funding and resources that are often found within public health systems. Potential terrorism and warfare bioregulators and toxins were given with an expected mortality of ≥50% were rated higher (+++) than agents with lower expected mortalities (21-49% = ++, and <21% = +). Toxins and bioregulators with higher rating (++) for morbidity were defined as such if clinical disease required hospitalization for treatment (including supportive care), and with lower rating (+) if outpatient treatment was required for most cases. Toxins and bioregulators received + to +++ for dissemination potential based on their likely methods of contamination of a large area by aerosol for respiratory exposure (+++), on contamination in quantities that could affect large populations (++), and sabotage on food and water supply (+). High level of intoxication by various routes is shown according of the kind of exposure: per oral route (+), respiratory route (++), or both (+++). Bioregulators and toxins also were ranked based on any special public health preparedness that was required including: stockpiling of therapeutics (+), enhanced surveillance and education (+), and improved laboratory diagnostics (+). Public fear associated with an agent and the potential mass civil disruptions that may be associated with even a few cases of disease were also considered (+ to +++).

The additional list of criteria for toxins and a table of toxins should be used for more detailed risk assessment and for comparison of the toxins on the list in which the lower the total number, the more dangerous toxin as terrorism or warfare agent, and with the purpose of making final decision on the list of toxins easier. In the additional list of criteria for toxins, one can list additional, less critical criterion: Ease of decontamination. For toxins, decontamination would be relatively unimportant but fungal toxins are extremely difficult to decontaminate and once an area is contaminated (infected), it can take years of hard cleaning to get rid of the toxins, especially if delivered with fungal spores.

For the evaluation of toxins, the toxicity should be interpreted with caution and certainly should include agents that are non-lethal but cause incapacitance. Lethality alone is not an appropriate criterion on which to base toxin's potential. For evaluation of toxins, the military preferred toxins that are primarily incapacitating to lethal toxins because the former have high operational impact and they represent a significant treat in the future. For example, Staphylococcal enterotoxin B (SEB), a so-called superantigen, is one the most potent agents for incapacitating at extremely low doses, but relatively high doses are required to kill. In this group of incapacitating toxins we can include trichotecene-mycotoxins, batrachotoxin and brevetoxin.

2.1 Criteria for Selection of Toxins and Bioregulators as Terrorism Agents

1. High level of morbidity: higher rating (++) if clinical disease requires hospitalization for treatment including supportive care and lower rating (+) if outpatient treatment is possible for most cases.
2. High level of mortality or incapacity: agents with an expected mortality of ≥50% were rated higher (+++), and with lower expected mortalities (21-49%=++, and <21%=+).
3. Stability in the environment after release (+)
4. Ease of production and transportation (+)
5. Likely methods for terrorism usage and high level of dissemination or contamination by aerosol for respiratory exposure (+++), contamination in quantities that could affect large populations (++), and dissemination potential for sabotage on food and water supply (+).
6. High toxicity or potency or low toxic dose: LD_{50} <0,000025 mg/kg (+++), LD_{50} from 0,000025 to 0,0025 mg/kg (++) and LD_{50} >0,0025 mg/kg (+).

7. High level of intoxication by variety route: per oral route (+), respiratory route (++), or both (+++).
8. Stockpiling of prophylactics and antidotal therapy (+)
9. Enhanced surveillance and education (+)
10. Difficult to diagnose or identify at the early stage or improved laboratory diagnostics (+).
11. Public perception: Public fear associated with an agent and the potential mass civil disruptions that may be associated with even a few cases of disease were also considered (+ to +++).

2.2 Criteria for Selection of Toxins and Bioregulators as Warfare Agents

1. Agents known to have been developed, produced, stockpiled or used as weapons (+).
2. High level of dissemination potential for contamination a large area by aerosol for inhalathory exposure and in military significant quantities that could affect large populations (+++), and dissemination potential for contamination of food and water supply (++).
3. High toxicity or potency or low toxic dose: LD_{50} <0,000025 mg/kg (+++), LD_{50} from 0,000025 to 0,0025 mg/kg (++) and LD_{50} >0,0025 mg/kg (+).
4. High level of morbidity: higher rating (++) if clinical disease requires hospitalization for treatment including supportive care and lower rating (+) if outpatient treatment is possible for most cases.
5. High level of intoxication by variety route: per oral route (+), respiratory route (++), or both (+++).
6. High level of mortality or incapacity: agents with an expected mortality of ≥50% were rated higher (+++), and with lower expected mortalities (21-49%=++, and <21%=+).
7. No effective prophylaxis and therapy commonly available and widely in use (+).
8. Stability in the environment (+).
9. Difficulty to diagnose/detect or identify at the early stage (+).
10. Ease of production and transportation (+).

2.3 Additional Criteria for Toxins

Toxicity
1= Lethal dose (LD_{50}) in the 10^{-9}g/kg range.
10= Lethal dose (LD_{50}) in the 10^{-3}g/kg range.
Level of incapacity or mortality
1= Predominately cause incapacitating.
10= Predominantly lethal.
Likely methods of dissemination
1= Toxin could be aerosolized and delivered to cover large areas for aerosol contamination (large- scale dissemination). Toxin could be used in sabotage for contamination food and water.
10= Toxin could not be aerosolized and delivered to cover large areas for aerosol contamination. Toxin could be difficult used in sabotage.
Stability in environment/storage
1= Extremely stable in storage and environment.

10= Unstable in environment or requires special storage conditions.

Onset

1= Minutes or hours to onset.

10= Hours or days to onset.

Ease of decontamination

1= Extremely difficult to decontaminate after a toxin aerosol attack.

10= Decontamination would be relatively unimportant and general decontamination procedures effectively destroy toxin.

Ease of production and transportation

1= Toxin can be ease produced in large quantities - low technology, low cost, widely available (fermentation).

10= Toxin that is very difficult to produce in weaponizable quantities - high cost, only available to specialized teams (solid phase synthesis of >100 amino acid polypeptides, advanced genetic manipulation).

3. Results and Discussion

Our opinion is that if a bioregulator or toxin satisfies a majority of the criteria, it should be recommended for inclusion on the list. As the list of bioregulators and toxins will be hard to define for purposes of the future negotiations of the States Parties of BTWC, this paper proposes five tables of enlisted bioregulators and toxins with important criteria on the basis of which a decision can be made to include in or exclude from a list of the toxins and the molecular agents (bioregulators). Rankings of potential bioregulators and toxins according to important criteria are shown in:

Table 1. Bioregulator assessment according to criteria for selecting bioregulators as warfare agents, and

Table 2. Bioregulator assessment according to criteria for selecting bioregulators as terrorism agents.

Table 3. Toxin assessment according to criteria for selecting toxins as warfare agents.

Table 4. Toxin assessment according to criteria for selecting toxins as terrorism agents.

Table 5. Toxin Risk Assessment (the lower the total number means the more dangerous the toxin as warfare agent).

4. Conclusion

It is very difficult to make a final decision on criteria and the final list of the molecular agents (bioregulators) for the needs of future Protocol to the BTWC based on these criteria. Therefore, this paper proposes that list and criteria for bioregulators be well studied and that an opinion by scientists and experts be obtained, because the list should be scientifically based. Although many bioregulators can be used to cause illness, they can truly threaten civilian populations on a large scale given the ability to produce and disseminate. If released upon a civilian population, these agents would pose the most significant challenge for public health and medical responses.

The above criteria for ranking potential toxins and bioregulators and listing of those of greatest public health concern could be used for determination of priority biological threat agents for national public health preparedness efforts for bioterrorism. Having a defined method for evaluating biological threat agents allows for a more objective

Table 1. Bioregulator assessment according to criteria for selecting bioregulators as warfare agents.

Bioregulator	(1) Weapo-nized (+)	(2) High level of dissemi-nation (+++)	(3) High tox-icity (+++)	(4) High morbi-dity (++)	(5) Intoxication by variety of route: per oral route, respiratory route, or both (+++)	(6) High level of incapacity/ mortality (+++)	(7) No effective prophylaxis and therapy (+)	(8) Stability in the environ-ment (+)	(9) Difficulty of detection/ identifi-cation (+)	(10) Ease of produc-tion (+)	Total (19)
1. Endorphins (Proopiomelanocortin – POMC; Pro-ACTH-Endorphin)	+	+++	+++	++	+++	+++	+	+	+	+	19
2. Substance P (SP) (Neurokinin)	+	+++	+++	++	+++	+++	+	+	+	+	19
3. Endothelins (ET-1, ET-2, ET-3) or Sarafotoxins (S6a, S6b)	+	+++	+++	++	+++	+++	+	+	+	+	19
4. Bradykinin (Kinin-9, Kallidin)	+	+++	+++	++	+++	+++	+	+	+	+	19
5. Vasopressin (VP)	+	+++	+++	++	+++	++	+	+	+	+	18
6. Angiotensins (I, II, III)	+	+++	+++	++	+++	++	+	+	+	+	18
7. Enkephalins	+	+++	+++	++	+++	++	+	+	+	+	18
8. Somatostatin (SS, SRIF)	+	+++	++	++	+++	++	+	+	+	+	17
9. Bombesin (BN)	+	+++	++	++	+++	++	+	+	+	+	17
10. Neurotensin	+	+++	++	++	+++	++	+	+	+	+	17
11. Oxytocin	+	+++	++	++	+++	++	+	+	+	+	17
12. Thyroliberins (Thyrotropin)	+	+++	++	++	+++	++	+	+	+	+	17
13. Histamine releasing factors (HRF): - Histamine release inhibiting factor (HRIF) - CTAP-3 Beta-Thromboglobulin (Beta-TG) - Neutrophil-activacting factor (NAF) - Stem cell factor (SCF)	+	+++	++	++	+++	++	+	+	+	+	17
14. Neurokinin A (NKA)/Substance K (SK)	-	+++	++	++	+++	++	+	-	+	+	15
15. Neurokinin B (NKB)/Neuromedin K	-	+++	++	++	+++	++	+	-	+	+	15
16. Neuropeptide Y (NPY)	-	+++	++	++	++	++	+	-	+	+	15

Table 2. Bioregulator assessment according to criteria for selecting bioregulators as terrorism agents.

Bioregulator	(1) High morbi-dity (++)	(2) High level of mortality/ incapacity (+++)	(3) Stability in the environ-ment (+)	(4) Ease of produc-tion (+)	(5) High level of dissemi-nation (+++)	(6) High tox-icity (+++)	(7) High level of intoxi-cation (+++)	(8) No effective prophylaxis and antidotal therapy (+)	(9) Enhanced surveillance and education (+)	(10) Difficulty of detection/ identifi-cation (+)	(11) Public percep-tion (+++)	Total (22)
1. Endorphins (Proopiomelanocortin – POMC; Pro-ACTH-Endorphin)	++	+++	+	+	+++	+++	+++	+	+	+	+++	22
2. Substance P (SP) (Neurokinin)	++	+++	+	+	+++	+++	+++	+	+	+	+++	22
3. Endothelins (ET-1, ET-2, ET-3) or Sarafotoxins (S6a, S6b)	++	+++	+	+	+++	+++	+++	+	+	+	+++	22
4. Bradykinin (Kinin-9, Kallidin)	++	+++	+	+	+++	+++	+++	+	+	+	+++	22
5. Vasopressin (VP)	++	+++	+	+	+++	+++	+++	+	+	+	+++	22
6. Angiotensins (I, II, III)	++	+++	+	+	+++	+++	+++	+	+	+	+++	22
7. Enkephalins	++	+++	+	+	+++	+++	+++	+	+	+	+++	22
8. Somatostatin (SS, SRIF)	++	++	+	+	+++	+++	+++	+	+	+	+++	21
9. Bombesin (BN)	++	++	+	+	+++	+++	+++	+	+	+	+++	21
10. Neurotensin	++	++	+	+	+++	+++	+++	-	+	+	+++	21
11. Oxytocin	++	++	+	+	+++	+++	+++	+	+	+	+++	21
12. Thyroliberins (Thyrotropin)	++	++	+	+	+++	+++	+++	+	+	+	+++	21
13. Histamine releasing factors (HRF): - Histamine release inhibiting factor (HRIF) - CTAP-3 Beta-Thromboglobulin (Beta-TG) - Neutrophil-activacting factor (NAF) - Stem cell factor (SCF)	++	++	+	+	+++	+++	+++	+	+	+	+++	21
14. Neurokinin A (NKA)/Substance K (SK)	++	++	-	+	+++	+++	+++	+	+	+	+++	20
15. Neurokinin B (NKB)/Neuromedin K	++	++	-	+	+++	+++	+++	+	+	+	+++	20
16. Neuropeptide Y (NPY)	++	++	-	+	+++	+++	+++	+	+	+	+++	20

Table 3. Toxin assessment according to criteria for selecting toxins as warfare agents.

Toxin	(1) Weapo-nized (+)	(2) High level of dissemi-nation (+++)	(3) High tox-icity (+++)	(4) High morbi-dity (++)	(5) Intoxication by variety of route: per oral route, respiratory route, or both (+++)	(6) High level of incapacity/ mortality (+++)	(7) No effective prophylaxis and therapy (+)	(8) Stability in the environ-ment (+)	(9) Difficulty of detection/ identifi-cation (+)	(10) Ease of produc-tion (+)	Total (19)
1. Botulinum toxin	+	+++	+++	++	+++	+++	+	+	+	+	19
2. Shigatoxin	+	+	++	++	+++	+++	+	+	+	+	16
3. Staphylococcal enterotoxin B (SEB)	+	+	++	++	+++	+++	+	+	+	+	16
4. Anatoxin A	+	+++	++	++	+++	++	+	-	+	+	16
5. Ricin	+	+	++	++	+++	+++	+	+	+	+	16
6. Modeccin	+	++	+	++	+++	+++	+	+	+	+	16
7. Saxitoxin	+	+	++	++	+++	+++	+	+	+	+	16
8. Tetrodotoxin	+	++	+	++	+++	++	+	+	+	+	15
9. Abrin	-	++	++	++	+++	++	+	+	+	+	15
10. Bungarotoxin	+	++	++	++	+++	++	+	-	+	-	14
11. Centruroides toxin	-	++	+	++	+++	++	+	+	+	+	14
12. Batrachotoxin	-	++	+	++	+++	++	+	+	+	+	14
13. Toxin of Clostridium perfrigens	+	+	++	++	+++	++	+	-	+	+	14
14. Trichotecene Mycotoxins (T2,DON,HT2)	+	+	+	++	+++	++	+	+	+	+	14
15. Ciguatoxin	-	+++	++	++	+++	++	+	-	+	-	14
16. Brevetoxin	-	++	+	++	+++	++	+	+	+	-	13
17. Palytoxin	-	+	+	++	+++	+++	+	+	+	-	13
18. Cyanginosin/Microcystin	-	+	+	++	+++	+++	+	-	+	-	12
19. Aflatoxin	+	++	+	+	++	+	+	+	+	+	12
20. Verrucologen	-	++	+	++	+++	++	+	-	+	-	12
21. Viscumin	-	++	+	++	++	++	+	+	+	-	12
22. Volkensin	-	+++	+	++	++	++	+	-	+	-	12

Table 4. Toxin assessment according to criteria for selecting toxins as terrorism agents.

Toxin	(1) High morbidity (++)	(2) High level of mortality/ incapacity (+++)	(3) Stability in the environment (+)	(4) Ease of production (+)	(5) High level of dissemination (+++)	(6) High toxicity (+++)	(7) High level of intoxication (+++)	(8) No effective prophylaxis and antidotal therapy (+)	(9) Enhanced surveillance and education (+)	(10) Difficulty of detection/ identification (+)	(11) Public perception (+++)	Total (22)
1. Botulinum toxin	++	+++	+	+	+++	+++	+++	+	+	+	+++	22
2. Ricin	++	+++	+	+	+++	++	++	+	+	+	+++	20
3. Staphylococcal enterotoxin B (SEB)	++	+++	+	+	++	++	+++	+	+	+	+++	20
4. Shigatoxin	++	+++	+	+	+	++	+++	+	+	+	+++	19
5. Saxitoxin	++	+++	+	+	++	++	++	+	+	+	+++	19
6. Abrin	++	++	+	+	+++	++	++	+	+	+	+++	19
7. Anatoxin A	++	++	-	+	++	++	+++	+	+	+	+++	18
8. Trichotecene Mycotoxins (T2,DON,HT2)	++	++	+	+	+++	+	++	+	+	+	+++	18
9. Modeccin	++	+++	+	+	++	+	++	+	+	+	+++	18
10. Tetrodotoxin	++	++	+	+	++	+	++	+	+	+	+++	17
11. Centruroides toxin	++	++	+	+	++	+	++	+	+	+	+++	17
12. Toxin of Clostridium perfrigens	++	++	-	+	+	++	++	+	+	+	+++	17
13. Ciguatoxin	++	++	-	-	++	++	++	+	+	+	+++	17
14. Brevetoxin	++	++	+	-	++	+	++	+	+	+	+++	17
15. Palytoxin	++	+++	+	-	++	+	++	+	+	+	+++	17
16. Cyanginosin/Microcystin	++	+++	-	-	++	+	++	+	+	+	+++	16
17. Batrachotoxin	++	++	-	-	++	++	++	-	+	+	+++	15
18. Bungarotoxin	++	++	-	-	++	++	++	-	+	+	+++	15
19. Aflatoxin	+	+	+	+	+++	+	+	+	+	+	+++	15
20. Viscumin	++	++	+	-	++	+	+	+	+	+	+++	15
21. Verrucologen	++	++	-	-	++	+	+	+	+	+	+++	14
22. Volkensin	++	++	-	-	++	+	+	+	+	+	+++	14

Table 5. Toxin Risk Assessment (the lower the total number means the more dangerous the toxin as warfare and terrorism agent).

Toxin	Toxicity	Onset	Level of incapacity/mortality	Likely methods of dissemination	Stability in the environment/ storage	Ease of deconta-mination	Ease of produc-tion	Total
1. Botulinum toxin	1	3	7	3	2	6	1	23
2. Shigatoxin	1	5	2	3	3	7	2	23
3. Staphylococcal enterotoxin B (SEB)	4	6	2	2	3	5	2	24
4. Trichotecene Mycotoxin (T2,DON,HT2)	7	4	7	2	1	2	2	25
5. Ricin	3	6	8	3	2	5	1	28
6. Modeccin	3	6	5	4	5	5	1	29
7. Abrin	2	6	5	5	5	5	1	29
8. Brevetoxin	6	6	2	4	2	3	8	31
9. Aflatoxin	4	8	5	5	5	1	3	31
10. Saxitoxin	3	2	8	3	3	7	5	31
11. Viscumin	3	6	5	5	6	6	1	32
12. Toxin of Cl. perfrigens	3	6	8	3	3	7	3	33
13. Centruroides toxin	3	4	6	5	2	5	8	33
14. Palytoxin	2	4	8	3	5	3	9	34
15. Tetrodotoxin	3	4	5	3	5	5	9	34
16. Verrucologen	3	7	6	5	6	6	3	36
17. Anatoxin A	5	1	6	7	6	8	3	36
18. Cyanginosin/Microcystin	5	2	5	3	7	7	8	37
19. Volkensin	4	5	7	6	7	5	4	38
20. Batrachotoxin	3	1	6	4	9	8	8	39
21. Bungarotoxin	3	4	6	5	8	7	8	41
22. Ciguatoxin	3	7	6	6	8	5	9	44

evaluation of newly emerging potential threat agents, as well as continued re-evaluation of established threat agents. Using this prioritization method can help focus public health activities related to bioterrorism detection and response and assist with the allocation of limited public health resources.

References

[1] Alibek, K.; Handelman, K., Biohazard: The chilling True Story of the Largest Covert Biological Weapons Program in the World-Trade from the Inside by the Man Who Ran It, Random House: New York, 1999.

[2] Rotz, D. L., Khan, S. A., Lillibridge, R. S., Ostroff, M. S. and Hughes, M. J., Priotritizing potential biological terrorism agents for public health preparedness in the united states: Overview of evaluation process and identified agents, National Center for Infectious Diseases, Centers for Disease Control and Prevention, Proceedings of CBMTS III, 2000.

[3] Hamilton M., Toxins and Mid-Spectrum Agents. ASA Newsletter, No. 66, 89-3, (1998).

[4] A FOA Briefing Book on Chemical Weapons, Theath, Effects and Protection. SIPRI, Oxford University Press, Oxford, (1987).

[5] Clark, K., The Chemical Weapons Convention: Chemical and Toxin Warfare Agents and Disarmament, Cranfield University, Royal Military College of Science. (1997).

[6] Therrien M. and Drouin J., Molecular determinants for cell specificity and glucocorticoid repression of the proopiomelanocortin gene. Annals of the New York Academy of Science, 1993, 680: 663-71.

[7] Grant J.A. et al, Histamine-releasing factors and inhibitory factors. International Arch. Allergy Appl. Immunology, 1991, 94: 141-3.

[8] Kuna P. et al, Further characterization of histamine releasing chemokines present in fractionated supernatants derived from human mononuclear cells. Clin. Experimental Allergy, 1996, 26(8): 926-33.

[9]Chen L.I. et al, The interaction of insulin and angiotensin II on the regulation of human neuroblastoma cell growth. Mol. Chem. Neuropathology, 1993, 18: 189-96.

[10] Clyne C.D. et al, Angiotensin II stimulates growth and steroidogenesis in zona fasciculata/reticularis cells from bovine adrenal cortex via the AT1 receptor subtype. Endocrinology, 1993, 132: 2206-12.

[11] Wolf G. and Neilson E.G., Angiotensin II as a hypertrophogenic cytokine for proximal tubular cells. Kidney International Suppl., 1993, 39: S100-7.

[12] Yamamoto Y. et al, Angiotensin III is a new chemoattractant for polymorphonuclear leukocytes. Biochemical and Biophysical Research Communications, 1993, 193: 1038-43.

[13] Rozengurt E., Neuropeptides as cellular growth factors: Role of multiple signaling pathways. European Journal of Clinical Investigation, 1991, 21: 123-34.

[14] Woll P.J. and Rozengurt E., Two classes of antagonist interact with receptors for the mitogenic neuropeptides bombesin, bradykinin, and vasopressin. Growth Factors, 1988, 1: 75-83.

[15] Burch R.M. and Kyle D.J., Recent developments in the understanding of bradykinin receptors. Life Sci., 1992, 50: 829-38.

[16] Hess J.F. et al, Cloning and pharmacological characterization of a human bradykinin (BK-2) receptor. Biochemical and Biophysical Research Communications, 1992, 184: 260-8.

[17] Leppaluoto J. and Ruskoaho H., Endothelin peptides: biological activities, cellular signaling and clinical significance. Ann. Med., 1992, 24: 153-61.

[18] Miller R.C. et al, Endothelins - from receptors to medicine Trends in Pharmacological Sciences, 1993, 14: 54-60.

[19] Moore, G. J., Designinig peptide mimetics. Trends in Pharmacological Sciences 15: 124-129 (1994).

[20] Geissler, E., Biological and toxin weapons today. SIPRI, Oxford University Press, Oxford, (1986).

[21] Daly, J. W., The chemistry of poisons in amphibian skin. Proc. Nat'l. Acad. Sci. USA. 92: 9-18 (1995).

[22] FM 3-9, Chemical Agents and their properties, Headquarters Department of the Army, (1994) Washington, DC., (1994).

Key words: toxins, bioregulators, warfare and terrorism agents.

Toxicogenomics and Proteomics
J.J. Valdes and J.W. Sekowski (Eds.)
IOS Press, 2004

Prospects on Immunoassays for the Detection of Pesticides in the Environment

Nabil A. MANSOUR and Ahmed S. ALASSUITY
Pesticide Chemistry and Toxicology
University of Alexandria
Alexandria, Egypt

Abstract. Challenges are many for the toxicologists and environmentalists, especially in the development of new toxicological parameters and high precision and sensitive technologies for environmental monitoring of hazardous and persistent pollutants. Advances in molecular biology have resulted in a dramatic increase in knowledge about gene structure and function, advancement in database of genetic sequence information, and very fruitful and advanced technologies for monitoring genetic sequences, variation and gene expression. These advances have achieved progresses in the correlation between the genome and the adverse effects of toxicants or the cellular modifying agents. Precise detection of environmental pollutants is of great importance to genotoxicologists. The analysis of organophosphorus (OP) pesticides and their derivatives using immunochemical methods is gaining acceptance as a simple, sensitive, selective and cost-effective screening of many samples. An indirect competitive enzyme-linked immunosorbent assays (ELISA) in the immobilized antigen format were developed for the OP insecticide chlorpyrifos (CPF), its metabolite trichloro-pyridinol (TCP), and the herbicide triclopyr (TRI). The parent molecule was derivatized by substitution of the chlorine in the 6-position of the pyridyl ring to obtain hapten-1 that was used to produce hapten-3 by hydrolysis. BSA conjugates for haptens 1, 2, and 3 were used to produce polyclonal *antisera*. At optimized ELISA conditions, the quantitative detection range in water for CPF was from 0.01-0.5 µg/ml with LOD of 0.002 ng/ml and an IC_{50} value of 30 ng/ml. While values were from 0.07-9.0 µg/ml, 0.02 µg/ml and 0.65 µg/ml for triclopyr, respectively. Comparison was carried out for the analysis of these three compounds in water between ELISA, fluoroimmunoassay and fiber optic immunosensor. The ELISA turned out to be a convenient tool for a preliminary screening for the three compounds in environmental samples.

1. Introduction

Pesticides are used to enhance agricultural production, to reduce pest and weed damage to crops and foodstuffs in storage and to control disease vectors in public health programs. In all these applications there is a risk that unwanted side effects on the environment or human health could, as the pesticides are inherently toxic to living organisms. *However, not all of the pesticides are highly toxic to humans. Proper choice of pesticides and their formulations and other proper precautions will reduce the health risk.*

There can be few aspects of pesticide research where it is more important to define objectives than in the case of environmental studies. Circumstances of pesticide use are tremendously varied, the aims of conservation differ enormously between countries and so do the aims of agriculture or public health. The ultimate goal is to avoid unacceptable effects on valued environmental components. The unacceptable effects are varying from country to country. They vary according to those who have to accept them. In fact, research should be concentrated towards looking at potential effects on all classes of non-target species. Such research consists essentially of: developing data; using it to adapt practices to minimize ill effects; and making value judgments between benefits and foreseen ill effects. It is the role primarily of the regional scientists with the local regulatory authorities for interpretation and adaptation of use recommendations.

Growing evidence indicates however, that trace amounts of pesticides are present on nonagricultural land, in the atmosphere, and in water, both in surface bodies and underground, far from the sites of pesticide application. Since pesticides are toxic by design, there is a natural concern on the impact of their presence in the environment on human health and environmental quality. In recent years, the adverse effects of certain pesticides to human and environmental health have become known. Although sporadic information on these adverse effects can be found in the literature since the early days of pesticide use, the impetus for the current awareness and concern for the adverse effects of pesticides probably had its beginning in the 1960s. Today, in addition to concern for the acute and chronic toxicity of pesticide chemicals, their potential as carcinogens, teratogens, and mutagens and their presence in all corners of the Earth, from the arctic icecap to the groundwater's used for drinking purposes, have led to questions about the wisdom of continued pesticide use.

These concerns on the potential threats of pesticides on human health and environmental quality must be viewed as serious. Pesticide use may have to be restricted to certain geographic locations or under certain agricultural practices. On the other hand, before any decisions are made to eliminate pesticide use for agricultural production purposes, it must be established that pesticides in unwanted locations is from pesticide use in agricultural production and not the result of other means of transport such as improper disposal of pesticides.

Reports indicated that the pesticide use in agriculture in Egypt increased until the early 1980s, coinciding with the growth and intensification in crop acres and a larger share of crop acres receiving pesticide treatments. Since the 1982 peak, the quantity of pesticide active ingredients used by farmers has fluctuated between 15 and 30 thousand metric tons. The annual fluctuations (with a remarkable drop in the 90s) have partially been in response to the government's national strategies of implementing IPM programs with emphasis on using alternatives to chemical pesticides such as natural products and/or the effect of newer products applied at lower rates [1-4].

On the international scale, organophosphates seem likely to continue to be the most important types of insecticides used in developing countries and countries in economic transition, and demand for them has been doubled in the last 10 years. Carbamate usage is also likely to increase substantially, while that of organochlorine was due to decrease considerably, since 1972 worldwide. On the other hand, there will be considerable increase in the use of synthetic pyrethroids and non-chemical types of pest control. The demands for both triazine and carbamate herbicides probably tripled by the late 90s. The most important class of fungicides continued to be the dithiocarbamates, which has increased in the last 10 years by four-fold.

Environmental pressures are impacting in many ways. They are leading to move stringent registration requirements for both new and old products and, regrettably, longer registration approval times and arbitary restrictions on product registration. The potential for such widescale environmental contamination had been discussed by several scientists shortly after DDT's introduction, but their concerns received little public attention until the publication of Rachel Carson's "Silent Spring" in 1962. Looking back, "Silent Spring" can be called the Birth cry of the environmental movement. This concern sparked the first Earth Day in 1970, the 1972 banning of DDT in the U.S. and the creation of the U.S. Environmental Protection Agency (EPA).

Because pesticides are applied to small or large areas of crops or water, they can pollute food, air, water and soil, so that the environmental effects of these chemicals can take many forms. Considerable care is needed to avoid the pollution of human food with pesticides, and this is achieved in developed countries by regular market basket surveys but in developing countries is still uncontrolled. On the other hand, some pesticides are persistent and can affect ecosystems for a number of years [5]. However, less persistent chemicals that are very toxic broad spectrum biocides can also have quite drastic and long-lasting environmental effects.

From environmental distribution to the genetic disease, for most pesticides the toxicologic paradigm is a dose-effect relationship. In these situations we have to deal with both acute and chronic toxicity, and in some cases with the phenomena of delayed polyneuropathy such as organophosphate induced delayed polyneuropathy (OPIDP). It is important to remember that the mode of action in acute toxicity is usually related in target and nontarget species, and that control of acute toxicity to man from nonpersistent insecticides is largely a matter of accident prevention. In the case of chronic toxicity, the modes of action are not well understood and the effects are long-term difficult to anticipate, and related to environmental fate. Although many problems exist in the definition of acute and chronic toxicity, it is apparent that they can be distinguished, in a general way by exposure, pharmacokinetic, and response parameters. Considerable understanding for such interactions could be achieved by, ways in which man is exposed to pesticides and the effects on man from the angle of the acute over-exposure of pesticide. These parameters should be studied carefully for each pesticide with special emphases on developing techniques for early prediction of pesticide effects on man. The effects of pesticides may be detected by measuring minor biochemical changes before adverse clinical health effects occur. There may be a threshold below which no effects can be observed (no-observed-adverse-effect level "NOAEL"). However, for pesticides that are suspected of causing cancer, the detection of early minor effects and the setting of NOAEL may not be relevant. The combination of toxicity vs. use rate is urgently required for developing safer compounds and safety concerns are required.

Pesticide weight, as a measure of pesticide use, has two particularly notable drawbacks when used for evaluating the potential for harm to human health and the environment. **First**, pesticide active ingredients vary widely in term of toxicity per unit of weight, irrespective of the scale used to measure toxicity. **Secondly**, weight does not account for the persistence of the pesticide in the environment. The longer a pesticide ingredient remains active in the environment, the more potential there is for it to come into contact with unintended species. Persistence varies widely among active ingredients, but many modern pesticides have half-lives (the typical measure of persistence) in the range of 10 to 100 days. In addition, other persistent pesticides would last in the environment for more than 10 years.

Challenges are many for the toxicologists and environmentalists, especially in the development of new toxicological parameters and high precision and accurate technologies for environmental monitoring of hazardous and persistent pollutants. The unprecedented advances in molecular biology during the last two decades have resulted in a dramatic increase in knowledge about gene structure and function, an immense database of genetic sequence information, and an impressive set of efficient new technologies for monitoring genetic sequences, genetic variation, and global function gene expression. These advances have led to a new sub-discipline of toxicology, toxicogenomics, "for the study of the relationship between the structure and activity of the genome and the adverse biological effects of exogenous agents". Likewise, there are powerful new methods for protein analysis or proteomics, "the study of the complement of proteins in the cell", and for analysis of cellular small molecules and metabolomics, the study of the cellular metabolites formed and degraded under genetic control [6].

The application of genomic technology to toxicology (toxicogenomics) has allowed the simultaneous identification of modified gene expression in response to a toxicant to be established for thousands of mammalian genes. This, together with the development of proteomics, metabolomics and our increasing understanding of individual human polymorphisms, will enable toxicologists in this century to identify those individuals at particular risk from specific toxins, pesticides and pharmaceutical drugs. However, these new opportunities will prove fruitless unless toxicologists address some of the major issues that presently confront their discipline. If anything, the new technologies impose a greater demand on toxicologists to exercise expert judgment on the meaning of their data, and to apply 'common sense' when balancing risks and benefits [7]. It is anticipated that, in the future, toxicologists will apply a range of genomic and proteomic techniques "Toxicoproteomics," to address issues in toxicity [8].

Assessment of the environmental hazard posed by the environmental components containing low to moderate levels of contaminants using standard analytical chemical methods is uncertain in several cases (HPLC, GLC, and GC/MS) due (in part) to a lack of information on contaminant bioavailability, the unknown interactive effects of contaminant mixtures, and the relative sensitivity of bioassay species. Transcriptional level (mRNA) toxicity assessments offer great advantages in terms of speed, cost and sample throughput. These advantages are currently offset by questions about the environmental relevance of molecular level responses. To this end a flow-through, high-density DNA hybridization array (genosensor) system specifically designed for environmental risk assessment was developed. The genosensor is based on highly regular microchannel glass wafers to which gene probes are covalently bound at discrete and addressable locations. The flow-through design enables hybridization and washing times to be reduced from approximately 18h to 20 min. The standard microscopic slide format facilitates data capture with most automated array readers and, thus high sample throughput (> 350 sample/h). Hardware development for molecular analysis is enabling very tractable means for analyzing RNA and DNA. These developments have underscored the need for further developmental work in probe design software, and the need to relate transcriptional level data to whole-organism toxicity indicators [9].

2. Pesticides and Persistent Organic Pollutants (POPs)

2.1. A POPs Prime: General view and status of pesticides

Every person alive today carries approximately 250 chemicals within her or his body, chemicals that did not exist prior to 1945. This concentration of chemicals within every human being on the face of the earth is called the "body burden," and it is our common legacy from the processes of development and industrialization.

World War II was a catalyst for the transformation from a carbohydrate-based economy to a petrochemical-based economy, as chemical substitutes began to be invented for goods restricted or made unavailable during the war. The economic boom that followed World War II supported the parallel boom in the invention and use of chemicals, many of which are associated with the convenience and flexibility of modern living. Environmental health advocates remind us that pesticides and herbicides have increased crop and livestock production, new drugs have curtailed or ameliorated many diseases, and plastics have found many uses within households around the world. All told, about 100,000 chemicals have entered into the market since 1945, and it is estimated that 75,000 of them remain in commercial use. The United States of America alone has increased its volume of synthetic chemicals one thousand-fold over the last 60 years.

These synthetic chemicals find their way into everything- soil, air, water and food. They are in the tissues of plants, animals and people. A startling fact about this increase of synthetic chemicals is that most remain untested for their safety in humans and other species. Today, only about 1.5% to 3% (about 1,200 to, 500 chemicals) has been tested to determine if they are carcinogenic. No one knows about the risks of cancer carried by the rest. Anecdotal evidence suggests a high correlation between many untested chemicals and cancer, as well as with many other diseases such as immune system dysfunction, reproductive failure and neurological problems. Moreover, chemical testing tends to study one chemical at a time, whereas real-life exposure is usually to a broad spectrum of chemicals that may interact or have additive effects.

Chemical testing is based on the idea that damage will occur after a certain level of exposure has been reached, and that exposure below these levels will cause no harm. Exposure studies therefore often start at elevated levels close to the point where cancer or DNA damage is expected, and measurements are done on adult laboratory animals. Risk assessment has to do with the likelihood of exposure for human populations who, because of their work, living situation or diet, risk exposure. Yet emerging science is beginning to indicate that fetal contamination, for example, can occur at very low levels which are currently not being tested for, and that the timing of exposure may also play a critical role in terms of possible effects.

However, the information on toxicity for some synthetic chemicals has been of sufficient concern to encourage a number of governments to ban or severely restrict their use. Many of these are pesticides or herbicides, or by-products or components of industrial processes. Among these are chemicals called "Persistent Organic Pollutants" or POPs. A proposed international agreement for the elimination or severe restriction of these chemicals a present and future task of the upcoming UN organizations negotiations. Many governments have already eliminated or severely restricted POPs on a national level. Since these particular chemicals are known for their ability to persist in the environment, to bioaccumulate in the food chain, and to travel long distances across national boundaries, the time is ripe for reaching an international agreement. POPs are a global problem that requires a global solution.

2.2. Definition of POPs

Persistent Organic Pollutants (POPs) are synthetic chemical substances composed of organic (carbon-based) chemical compounds and mixtures. POPs are products and by-products of human industry that are of relatively recent origin. In the early decades of this century, pollutants with these harmful properties were virtually non-existent in the environment and in food.

Now, ordinary food supplies in most regions of the world, especially fish, meat and dairy products tend to be contaminated by POPs. Both people and wildlife, everywhere in the world, carry body burdens of POPs at or near levels that can, and often do, cause injury to human health and to entire ecosystems. Because they generally have low water solubility and high lipid solubility, they tend to bioaccumulate in fatty tissues of living organisms. In the environment, concentrations of these substances can magnify by factors of many thousands as they move up the food chain.

What distinguishes POPs from other such substances is that they can travel in the environment to regions far from their original source, and then can concentrate in flora and fauna to levels with the potential to injure human health and/or the environment. POPs are persistent in the environment. This means that they are substances that resist photolytic, chemical and biological degradation. They are generally semi-volatile. Persistent substances with this property tend to enter the air, travel long distances on air currents and then return to earth. They are also subject to global distillation (i.e., migration from warmer to colder regions).

POPs are also highly toxic, having the potential to injure human health and the environment at very low concentrations. In some cases, POPs at concentrations of only one or a few molecules per cell can attach to intercellular receptor sites and trigger a cascade of potentially harmful effects.

2.3. Efforts to phase out POPs

After several meetings, UNEP experts and ministers from 110 countries met in Washington D.C., in November 1995 for a final negotiating session to adopt a Global Program of Action for the Protection of the Marine Environment from Land-Based Activities. The ministers agreed on the need to develop a "global, legally binding instrument for the reduction and/or elimination of emissions, discharges and where appropriate, the manufacture and use" of 12 of the most persistent, bioaccumulative organochlorine chemicals that have been found to pollute the marine environment.

During 1996 and 1997, intergovernmental organizations including the Intergovernmental Forum on Chemical Safety (IFCS), the World Health Organization (WHO) and UNEP refined and elaborated on this decision and agreed to create an Intergovernmental Negotiating Committee (INC) to establish the terms of this instrument. The committee's first meeting held on June 29, 1998 in Montreal, Canada. Negotiators were asked to mandate action on a short list of 12 POPs, sometimes called the "dirty dozen." They are: dioxins, furans, polychlorinated biphenyl's (PCBs), DDT, chlordane, heptachlor, toxaphene hexachlorobenzene, aldrin, dieldrin, endrin, and mirex. Among the twelve POP substances there are four unintentionally generated by-products, generated by human activities, they are: Polychlorinated dibenzo-*p*-dioxins and dibenzofurans (PCDD/PCDF), Hexachlorobenzene (HCB) and polychlorinated biphenyls (PCB). While HCB is a single chemical compound, PCDDs have 75 different theoretical combinations (Congeners), PCDFs have 135 congeners, and PCBs have 209. It should be noted that the toxicity and also the resistance against destruction (persistence) varies widely among the congeners. Only 7 of the 75 congeners of PCDDs and 10 of the 135 possible congeners of PCDFs are thought to have dioxin-like toxicity.

Action on POPs will require a great deal of concentrated effort. Many countries have no way of knowing the levels of contamination in their countries and need the technology and training to measure exposure levels. Many countries want to learn about and have access to alternatives to POPs that will help them maintain a constant food supply or a means to control disease vectors. Stockpiles need to be identified, and waste disposal methods developed which will do no further harm to the environment. Funds and technology transfer will be necessary for countries to develop the means to restrict or eliminate the use of POPs.

2.4. Toxicological Parameters of POPs

Damage caused to humans and other species by POPs is well-documented and includes the pathologies of cancer and tumors at multiple sites, reproductive disorders, neurobehavioral impairment including learning disorders, immune system dysfunction, lack of development in various body systems such as the reproductive system, immune system, endocrine system, and neurological systems, adverse effects to the adrenal glands, the liver and the kidneys, heart disease, cerebrovascular disease, stillbirths, and behavioral changes such as fatigue, depression, personality changes, tremors, convulsions and hyperexcitability.

Some of these effects, it has been postulated, are caused by the fact that many POPs can act as endocrine disrupters. Endocrine disrupters are chemicals that can act as false hormones within the body. Hormones are the substances that turn on or turn off various mechanisms that trigger development. Since bodies cannot recognize the difference between natural hormones and false or 'xeno- hormones,' these chemicals can alter in alarming ways the functioning of a human body or the bodies of other species.

The greatest damage occurs during pregnancy, when these chemicals mimic or block the miraculously delicate signals that the mother's hormonal system sends to the developing fetus to guide its development. According to some recent scientific studies by Colborn [10], DeVito [11], Jacobson and the EPA [12], as the child develops, endocrine disruption in the womb and through breast milk may result in cancer, endometriosis, learning disorders, behavioral disorders, immune and neurological disorders and a wide range of other problematic conditions like low sperm count, low IQ, genital malformations and infertility.

The more scientists learn about endocrine disrupting chemicals, the more troubled they have become, First, there appears to be no minimum dose at which these chemicals are safe for a developing fetus, Since the usual governmental mechanism of "risk assessment" which gives industry the right to expose people to toxic chemicals assumes that there are safe doses, this new finding of no safe minimum dose may mean, as one scientist put it, "the end of risk assessment as we know it". Second, many of these endocrine-disrupting chemicals have different effects on the developing fetus at different "developmental windows" and at different dosages. A smaller dose at one window may have a completely different effect than a larger dose at another window. Third, the impact of many of these endocrine-disrupting chemicals appears in many instances to be additive or even synergistic. To properly evaluate their full health effects, scientist would have to test for all the mixtures that developing fetuses are actually exposed to at all the different times they might be exposed. Scientists have only begun to undertake this task, made especially complex by the fact that many effects may not appear until offspring reach puberty. Fourth, people are already carrying loads of many of these chemicals, at levels at which there are known health effects in either animals or human. People do not have "room" for additional exposures.

3. Environmental Monitoring of Pesticides by Immunochemical Techniques

Due to the widespread use of pesticides, there is a growing concern over the environmental contamination caused by their residues, which demands, adequate monitoring. The analysis of pesticides and their derivatives using immunochemical methods is also, gaining acceptance as a simple, cost-effective screening of many samples prior to confirmatory chromatographic techniques. [13, 14].

Immunochemistry has broad applications for a wide variety of environmental contaminants. However, the potential for applying immunochemical methods to environmental measurements is beginning to be realized. Immunochemical methods are based on specific antibodies combining with their target analyte(s). Many specific antibodies have been produced for targets of environmental and human health concern. Such antibodies can be configured into various analytical methods. The most popular immunochemical technique in environmental analyses today is immunoassay.

Immunoassays have been shown to detect and quantify many compounds of environmental interest such as pesticides, industrial chemicals, and products of xenobiotic metabolism. Among the most important advantages of immunoassays are their speed, sensitivity, selectivity and cost-effectiveness.

Immunoassays can be designed as rapid field-portable, semi-quantitative methods or as standard quantitative laboratory procedures. They are well suited for the analysis of large numbers of samples and often obviate lengthy sample preparations. Immunoassays can be used as screening methods to identify samples needing further analysis by classical analytical methods. Immunoassays are especially applicable in situations where analysis by conventional methods is either impossible or is prohibitively expensive.

Environmental immunoassays have broad applications for monitoring studies. The EPA has used immunoassay methods for monitoring groundwater and cleanup activities at hazardous waste sites. Immunoassays can also be used as field screening tools to confirm the absence and or presence of particular contaminants or classes of contaminants for special survey.

In addition to detection methods, other immunochemical procedures can be used for environmental analysis. Immunoaffinity techniques now used extensively in pharmaceutical and biotechnology applications can be adapted to extract and cleanup environmental samples. Selective and sensitive sample collection systems such as air and personal exposure monitors can be designed based on the principal of immunoaffinity.

The Trans Boundary Diagnostic Analysis for the Mediterranean Basin identified the following as the main problems that affect water quality use: high load of nutrients and eutrophication; contamination with hazardous substances including POPs and oils; microbiological contamination; contamination with substances causing hetero tropic growth and oxygen depletion and competition for available water. The activities contributions significantly to these problems are human activities, agriculture, rivers and drainage discharges, industry and tourism. Marine pollution from land-based sources and activities has long been recognized as a major problem in the marine environment. It has been estimated that approximately 80% of the total pollution, especially in the "hot spots", of the Mediterranean Sea is generated by land-based sources and activities. One of the responsibilities, of these countries is to collaborate to develop and adopt the pesticide and persistant chemicals monitoring new technologies.

Methods and immunologic reagents have been developed for various pesticides (organophosphate insecticides metabolites and herbicides). Additional methods are under development for POPs and synthetic pyrethroid insecticides.

4. Development of Immunoassays for Pesticides

The need to evaluate the risk to the environment from the use of chemicals has been a significant part of the regulation of pesticides for many years. There has been an increased awareness and concern from the public and regulatory authorities regarding the potential for pesticides to contaminate air, soil and water sources. This pressure has resulted in the evaluation of different analytical methods and detection techniques in an effort to lower detection limits and improve confirmation procedures for pesticides, especially in water. A major risk is environmental contamination, especially translocation within the environment where pesticides may enter both food chains and natural water systems. Factors to be considered in this regard are persistence in the environment and potential for bioaccumulation judged by the most precise and accurate analytical procedures.

Organophosphorus pesticides (OPs) are a structurally diverse group of chemicals, and OP pesticides may be classified based on any number of structural similarities and differences. The reactivity of OP compounds varies depending upon the chemical structure. This is very important in our present study for developing the immunoassay technique for the phosphorothioate compound "Chlorpyrifos". Electrophilicity of the OP is crucial, in general, for the biological actions of OP compounds and, in particular, for developing immunochemical analytical methods. OP compounds that have a double bond between P and O are highly electrophilic at the P atom and are highly reactive. Groups that enhance the reactivity of the P are nitro, cyano, halogen, ketone, and carboxylic ester. Deactivating groups include hydroxyl and carboxylic acid [15, 16, 17, 18]. These features have to be considered in developing conjugates for immunoassay analysis for these compounds.

Chlorpyrifos "CPF" (O,O-diethyl O-3,5,6-trichloro-2-pyridyl phosphorothioate) an organophosphate insecticide, is a broad-spectrum insecticide with insecticidal activity against many insect and arthropod pests. Chlorpyrifos has been successfully utilized to combat insect and arthropod pests threatening the production of food and fiber and the maintenance of human health, and is registered for the control of different insect pests allover the world.

Because of the importance of CPF in practical applications around the world, an immunoassay technique for determination of CPF was developed as a model for most structurally related compounds. About the time chromatographic methods were being developed for the analysis of pesticide residues in food and the environmental components during the 1960's, immunoassays were catching on for the analysis of a wide array of analytes in the clinical chemistry laboratory. From their inception, chromatographic methods for multiresidue methods were concerned only with those pesticides that were considered to be a threat to human health, the insecticides, which at that time were predominantly the organochlorines [19]. All traditional techniques placing constraints on the pesticide analytical chemist because of several reasons; lengthy, tedious and labor-intensive or required very expensive equipment. In the late 1980s a growing concern by scientific researchers and particular authorities about accurate and nonexpensive analysis of chemical pesticides, the information gathering arms of authorities and scientists, to determine whether the development and implementation of newer pesticide residue analytical technologies would rectify the perceived problem. During this period the issue of further developing immunological techniques for pesticide residue monitoring was addressed, along with other technologies, such as supercritical fluid extractions (SFE), robotics, and biosensors.

Immunoanalysis is recognized as a major analytical method applicable to numerous analytical needs including detection and quantitation of drugs in body fluids and chemicals in environmental samples (e.g. rivers, underground water or soil extracts). Enzyme-Linked Immuno Sorbent Assays (ELISA) is the dominant format used at present time. Biosensors are analytical devices (to measure signals) that utilize biological materials (e.g. enzymes, are the most widely used proteins, receptors, antibodies, DNA) as sensing elements coupled to transducers (e.g. fiber optic/silicon detector, electrodes and ion selective field transistors "ISFT", surface plasmon resonance (SPR) and acoustic sensors). The biological material is immobilized directly on the transducer or on another matrix, which is brought in close proximity to the transducer. The technology is based on converting the energy produced from the binding reaction of the target chemical with the biological material to a measurable electrical signal that correlates with the concentration of the chemical.

4.1. Preparation of haptens

In this study three different haptens have been used Fig. (1). O,O-Diethyl O-[3,5-Dichloro-6-[(2-carboxyethyl) thio]-2-pyridyl] Phosphorothioate (hapten 1) has been synthesized according to the method described by Manclús *et al.*[20]. 3-(3,5-Dichloro-6-hydroxy-2-pyridyl) thiopropanoic acid (hapten 3) has been synthesized according to the method described by Manclús & Montoya [21] with certain modifications. Triclopyr a technical grade herbicide. [(3,5,6-Trichloro-2-pyridyl)oxy] acetic acid (hapten 2) was purchased from Chem. Service Co. (USA).

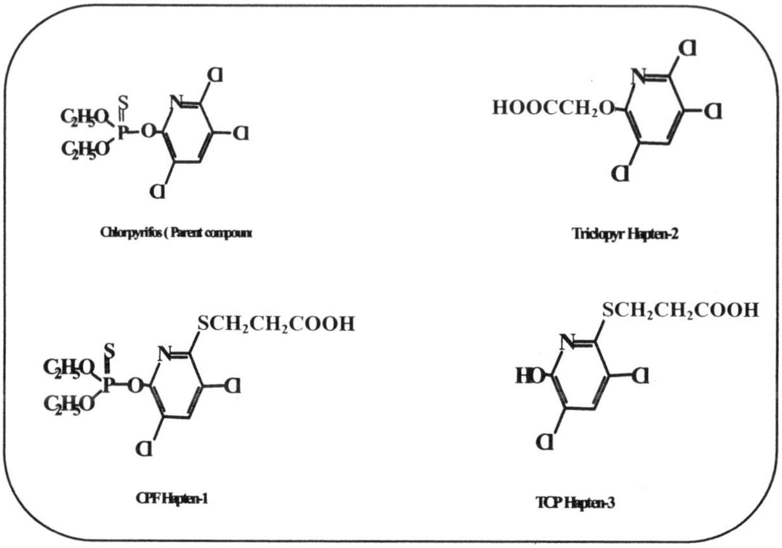

Fig. (1): Chemical structure of Hapten 1, 2 and 3.

Preparation of hapten-1

To a solution of 3-mercaptopropanoic acid (1.06g, "10 mmol") in 50 ml of absolute ethanol 2 equiv. of KOH (1.42g), was added and heated until dissolved. Then, chlorpyrifos (technical grade, 3.51g, "10 mmol") dissolved in 50 ml of absolute ethanol was added. After reflux for 1.5 h, the reaction mixture was filtered and the solvent was removed under reduced pressure. To the residue 50ml of 5% $NaHCO_3$ was added, then washed with hexane (3 x 50 ml). The aqueous layer was acidified to pH 4.0 and extracted with dichloromethane (3 x 50 ml). The extract was dried over Na_2SO_4 and concentrated. The residue was subjected to column chromatography [hexane/tetrahydrofuran (THF)/acetic acid 75:25:1]. Fractions showing only one spot on TLC (R_f, 0.41, with the same solvent mixture, compared with the standard starting materials) were pooled and concentrated to provide 1.05 ± 0.05 g of hapten-1 (25 %), which solidified on standing: mp 124-125°C; ^1H NMR ($CDCl_3$) & 7.76 (S, 1H, ArH), 4.37 (q + q, 4H, 2 CH_2O), 3.45 (t, 2H, SCH_2), 2.97 (t, 2H, CH_2COO), 1.42 (t, 6H, 2 CH_3).

Preparation of hapten-3

This hapten came from a chlorpyrifos hapten (*O,O-diethyl O-[*3,5-dichloro-6-[(2-carboxyethyl)thio]-2-pyridyl] phosphorothioate) previously synthesized by direct substitution of the chlorine in the 6-position of the pyridyl ring chlorpyrifos by a 3-

mercaptopropanoic acid as a spacer arm [20]. This hapten was prepared by hydrolysis of the thiophosphate ester as follow: To a solution of *O*, *O* - diethyl *O* - [3,5-dichloro -6- [(2-carboxyethyl)thio]-2-pyridyl] phosphorothioate, 250 mg in 3 ml THF, 5 ml 1 M NaOH was added, and the mixture was refluxed for 1.5 h. After adding distilled water (30 ml), the solution was acidified to pH 3.0 with 2 M HCl and extracted with ethyl acetate (EtAc; 3 x 30 mL). The combined organic extracts were dried over anhydrous sodium sulfate, and the solvent evaporated to give hapten-3 (85 %) as a white solid. ^{1}HNMR (dimethyl sulfoxide d6) & 7.92 (S, 1H, ArH), 3.26 (t, 2H, SCH_2), 2.66 (t, 2H, CH_2-COO).

Hapten (hapten-3) was synthesized for the trichloro-pyridinol (TCP), the main metabolite of CPF in the environment. Regarding its immunogenic structure, TCP is a small, simple analyte consisting of an aromatic ring with substituents. It was taken in our considerations that, an appropriate hapten design should preserve as many ring substituents as possible and produce minor modifications in the ring electric distribution as a consequence of spacer attachment.

Chemical properties of TCP molecule are determined by the ring substituents and the heteroatom, being the ionizable hydroxyl, and the active chlorine atom in the 6-position of ring putative sites for spacer attachment. Consequently, TCP hapten was prepared by splitting (hydrolysis) and removing the thiophosphate moiety of hapten-1 previously synthesized for the parent CPF insecticide Fig. (2). This hapten derivatization was accomplished by substitution of the chlorine in the 6-position with mercapto propinoic acid as previously described. The only structural modification introduced in this hapten was the spacer coupling as a thio-ether linkage instead of one of the chlorine substituents. This way, the chemical properties of TCP may be well preserved as suggested by the successful use of similar thio-ether linkage for spacer coupling to synthesize triazine herbicide haptens [22].

4.2. Preparation of Protein – Hapten Conjugates

All haptens used in this study contained a carboxylic group and were conjugated covalently to proteins by N-hydroxysuccinimide (NHS) active ester method [23], with slight modifications. Carboxylic acid hapten (0.20 mmol) was dissolved in 1.0 ml of dry dimethyl formamide (DMF) with hapten equimolar N-hydroxysuccinimide and a 10% molar excess of dicyclohexylcarbodiimide (DCC). After 3.5 h of stirring at 22°C, the precipitated dicyclohexylurea was removed by centrifugation, and the DMF supernatant was added to BSA or casein protein solution. The protein (50 mg) was dissolved in 5.00 ml of H_2O_2 and 1.05 ml of DMF was added slowly with vigorous stirring. The reaction mixture was stirred gently at 4°C for 22 h to complete the conjugation, then dialyzed exhaustively against 50 mM phosphate buffer saline pH, 7.4 (PBS), 24h & 4 changes with PBS.

The above modified method has been used to link all haptens to carrier proteins (BSA or casein) except for hapten 2-casein conjugate. Another modification was performed on the above method [21] as follows: Hapten [25 µmol (6.7 mg) in appropriate volume of DMF (200 µl) to bring the final concentration of hapten to (~25 mM)] was activated by incubation for 2 h at room temperature with a 50% molar excess of NHS (~ 4.3 mg) and DCC (~ 7.7 mg). The mixture was centrifuged. The supernatant was added to 5 ml solution of 10 mg/ml casein in 0.2 M borate buffer (pH 9.0) over a 10 min period and with vigorous stirring. The activated ester mixture diluted in the volume of DMF to bring the solution to 20% DMF. The initial hapten to protein molar ratio in the mixture was 50:1. Finally, the mixture was stirred at room temperature for 2h and dialyzed against PBS. Conjugate formation was confirmed spectrophotometrically and by elemental analysis.

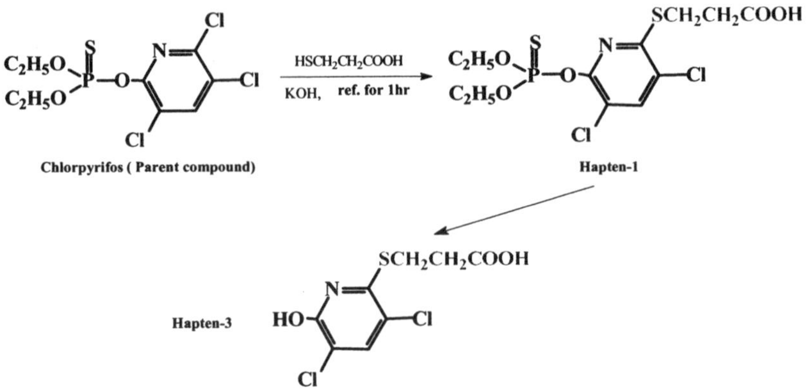

Fig. (2) : Preparation scheme of haptens (Hapten-1 & Hapten-3)

Two different types of conjugates were prepared, immunogenic conjugates (used for immunizing lab animals) and coating conjugates (used in Heterogenous ELISA format). Heterogenous ELISA format means using a different antigen than the one used in the immunization (immunizing conjugate) in coating the ELISA plates. Bovine serum albumin (BSA) was used as carrier protein in preparing immunogenic conjugates, while casein used in preparing coating conjugates. Conjugates formation was confirmed spectrophotometrically. UV-vis spectra showed qualitative differences between carrier proteins and conjugates in the region of maximum absorbance of haptens. BSA and casein as a protein have a maximum absorbance at 280nm. After coupling these proteins with hapten, a significant shift in the maximum absorbance. The UV-vis spectra of conjugates showed a maximum absorbencies at 315, 320 and 295 nm, for hapten-1 , hapten-2, and hapten-3, respectively.

Quantitative elemental analysis data of conjugates showed a significant chlorine percentage due to the linkage of carrier proteins to haptens (Table 1). The chlorine percentage varied from (4.37 to 9.9), (4.23 to 4.55), and 5.04 for hapten 1,2 and 3, respectively. All haptens used in this study contain two or three chlorine atoms, since proteins in general do not contain chlorine, percent of chlorine has been used as another method to confirm the linkage between hapten and carrier proteins. Quantitative elemental analysis data has been used to calculate the hapten to protein molar ratio (H/P ratio). There is a good correlation between the chlorine percentage and the number of haptens attached to each molecule of the protein. These calculations were therefore considered to be a good parameter in confirming the number of haptens per protein molecule. For hapten-1, 2, and 3 the H/P ratio is a good indicator for the conjugate hapten.

Mathematical calculations of chlorine percentage and the expected H/P ratio was formulated by blotting the percent of chlorine vs. H/P ratio. A straight line obtained and used in calculating the H/P ratio by using quantitative elemental analysis data (% Cl). The calculated H/P molar ratios of the prepared conjugates were 77, 45, and 34 for hapten 1, 2, and 3 , respectively.

Table 1: Quantitative Elemental Analysis of Chlorpyrifos (CPF), Triclopyr (Tri), and Trichloropyridinol (TCP) [Haptens*] Conjugated with Bovine Serum Albumin [Carrier Protein]

Immunogens**		Element (%)			
		C	H	N	Cl
CPF-BSA	Pellet	42.53	4.93	7.59	9.90
	Supernatant	48.38	6.49	13.73	4.37
Tri-BSA	Pellet	52.22	7.15	13.84	4.55
	Supernatant	49.05	6.53	13.60	4.23
TCP-BSA	Pellet	---	---	---	---
	Supernatant	---	---	---	5.04
BSA***		49.78	7.02	14.86	0.00

* *Hapten: small molecule that can't generate immune response.*
** *Hapten-BSA complex "Immunogen".*
*** *BSA used as a carrier protein.*

4.3. Immunoresponse to Conjugates

Three different animal species were used in this study rabbit; mouse, and chicken. Rabbits and mice were immunized according to the Current Protocols in Immunology [24] for the preparation of polyclonal antibodies. Immunization of the experimental chickens was carried out in Aves Labs Inc. Tigard, Oregon USA. The purified IgY from the egg yolk of the immune and non-immune chickens were used [25].

To test suitability of the synthesized immunizing haptens to elicit an appropriate antibody response to chlorpyrifos and its main metabolite TCP, the three different lab animal species were immunized with each of the BSA-hapten conjugates. After the fifth injection (the 4th boosting) blood was collected and sera prepared and subsequently characterized for the presence of antibodies recognizing the conjugated immunizing haptens "Serum-titer" (serum dilution giving three times the background absorbance). Serum titers were determined by indirect ELISA using casein-hapten conjugates (coating conjugates 25 μg/ml). All sera showed high levels of polyclonal antibodies recognizing each respective casein-hapten conjugates, with titer ranging from 1:400,000 to 1:500,000 for all three haptens, sera from Mouse and Rabbit (Fig 3).

Antibodies from chicken were purified from the eggs of the immunizing laying hens. The titer of chicken antibodies for CPF was 0.078 μg/ml, using the same ELISA format.

4.4. Immunochemical Techniques

4.4.1. Enzyme-Linked Immunosorbent Assays (ELISA)

ELISA has been used in this study either to determine the titer of the sera or in quantitative detection of the analytes using the competitive ELISA format.

ELISA experiments in the immobilized antigen format were performed [24] with certain modification. 96-well microtiter plates were coated with appropriate Analyte-Casein conjugates which had been dissolved in carbonate buffer (0.1M, pH 9.6), and then blocked with blocking buffer (50 mM TBS "tris-buffered saline", pH 8.0 + 10% fat-free dry milk + 0.1 % sodium azide). Sample or standard solutions and the antisera solutions all in blocking buffer were dispensed into the wells sequentially. After incubation, wells were treated with goat anti-animal (rabbit or mouse) IgG (whole molecule) alkaline phosphatase conjugate (1:500 in TBS + 0.5 % tween 20). Enzyme activity was then measured using *p*-nitrophenyl phosphate as substrate at 405 nm. To construct a standard curve competitive indirect ELISA was used as above, plus incubation of antibodies with serial dilutions of the analyte.

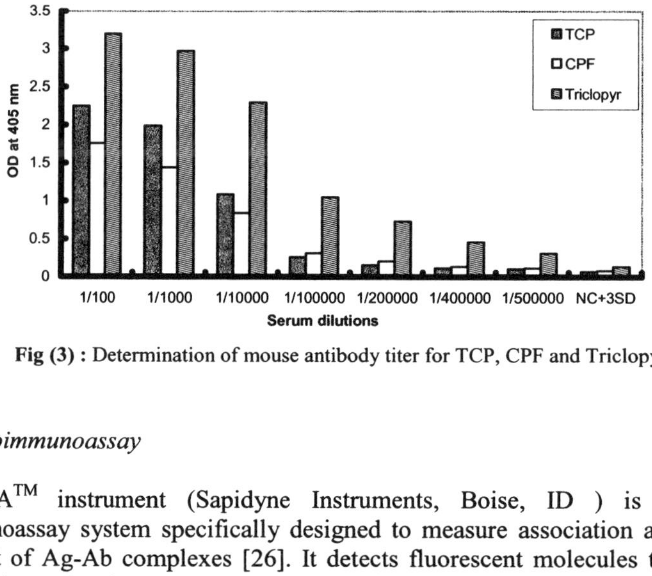

Fig (3) : Determination of mouse antibody titer for TCP, CPF and Triclopyr

4.4.2. Fluoroimmunoassay

The KinExA™ instrument (Sapidyne Instruments, Boise, ID) is an automated fluoroimmunoassay system specifically designed to measure association and dissociation rate constant of Ag-Ab complexes [26]. It detects fluorescent molecules that bind to the surface of polymer beads. A measured quantity of beads coated with Ag or Ab is introduced into a small capillary flow cell and retained on screen, forming a cylindrical packed bead bed approximately 3mm long in diameter. Ab (or Ag) flowing through the packed beads bed binds Ag (or Ab) attached to the beads. The portion of the capillary flow cell containing the bead bed is embedded in a lens and positioned between a light source and a reflector. This arrangement increases the efficiency of both fluorescent excitation and collection of fluorescent emissions.

The fluidics system accommodates up to 13 samples, and has two syringe pumps that operate under negative pressure to draw coated beads, buffer, samples and the fluorescent-conjugated reagent through the flow cell. A peristaltic pump back flushes the used beads out of the flow cell when a measurement is completed. The pumps and valve are activated and controlled by a PC-compatible computer, which controls the sequence of sampling, sample volumes, flow rates and number of replicate measurements [27].

The assay was able to detect as low as 0.0002 µg/ml of the analyte of interest. An inhibition curve was constructed (Fig. 4) for the herbicide triclopyr and it was linear over the range from 0.001 to 0.1 µg/ml of the analyte.

4.4.3. Fiber Optic Immunosensor (FOB)

The analyate 2000 system is a multi-channel FOB that is capable of performing multiple diverse immunoassays. Briefly, the system is operated as a flow fluorometer for detection of fluorescence generated by evanescent excitation on the surface of the tapered optical fibers. It uses four integrated circuit "daughter" cards containing a 5mW, 635-nm laser diode for fluorescence excitation, a photodetector, and a fiber optic coupler for attaching the fiber bundle jumper. The jumper is made of a fiber optic cable with a 200-µm center silica fiber carrying excitation light form the red laser diode to the tapered fiber and a cable that has six plastic fibers (250-µm core diameter) surrounding the center fiber, which collects and transmits fluorescence to the detector. Each card is coupled to an optical fiber using a fiber bundle jumper to provide for simultaneous monitoring of four immunoassays. Use of high numerical aperture (NA = 0.47) for the collection fibers ensures maximum fluorescence collection and detection. The sensing fiber is prepared by covalently coating the tapered optical fiber with antigen. This is used to capture Ab, which is subsequently detected using CY5-Ab (FluoroLink™, as a monofunctional Dye) [28].

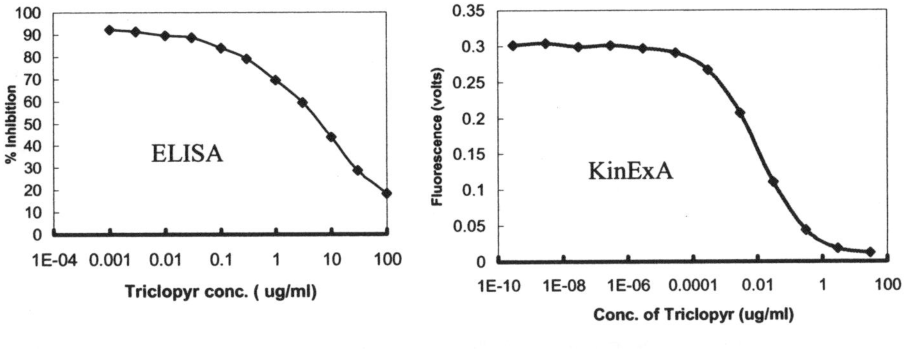

Fig.(4) :Inhibition curve for Triclopyr

The fluorescent signal was generated by sequential binding of Ab and then CY5-Ab to the Ag-coated fiber. It was reasoned that buffers used in affinity purification of Abs should be able to dissociate the bound Ab from the fiber, thus regenerating the fiber for reuse. Dissociation of the bound Abs was reflected in the rapid loss of fluorescence signal following wash with regenerating buffer. More than 95% of the regeneration (removal of CY5-Ab) occurred using 0.5M TEA, pH 11.5. Of the regenerating solutions used glycine buffer, pH 2.5 was the least effective, reducing fluorescence by only 20%. Increasing the concentration of buffer (Glycine / HCl) had a negligible effect on the dissociation.

The basic buffer 0.1M TEA (pH 11.5), was most efficient reducing the fluorescent signal by 75%. Higher concentrations improved the dissociation significantly. Accordingly, 0.5M TEA (pH 11.5) was used for regeneration of fibers in all subsequent assays. To investigate whether regenerated fibers produce accurate results, Ag-coated fibers were purfused sequentially with PBS-casein containing Ab and CY5-Ab to obtain the maximum fluorescent signal. Fibers were regenerated using 0.5M TEA, pH 11.5 buffer, and a new baseline was established. The assay was repeated on the same fiber after each regeneration the amplitude of signal from 10 regeneration did not change the fluorescent signal Fig. (5). After each fiber regeneration, the baseline fluorescent became higher. Repeated use and regeneration cycles resulted in a steady rise of the baseline fluorescence without significant change in the fluorescence signal. This was due to the incomplete dissociation of bound Abs from fiber. On average, each fiber was used for 15 measurements before the baseline become too high and variance increased above 10%.

5. Conclusions

The highly sensitive and selective ELISA for chlorpyrifos, its metabolite trichloro-pyridinol and the herbicide triclopyr gave advantage for reasonable, rapid and inexpensive analysis for quite large numbers of samples for environmental monitoring programs. These immunoassays may also provide a useful analytical tool for investigations on the biochemical mode of action and metabolism of these compounds. Immunoanalysis has achieved progresses in the last 3 decades noticeable and remarkable propagation in the last ten years. Table (2) represent several examples of immunoassay method for several fungicides, herbicides and insecticides monitoring in the environmental components.

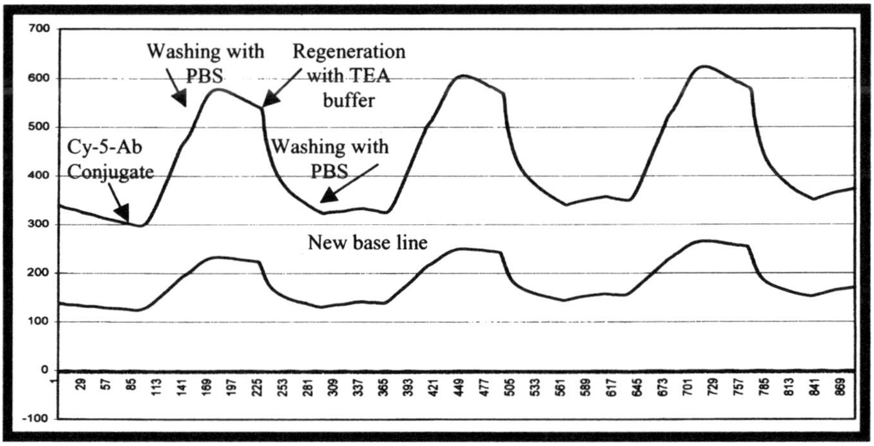

Fig.(5): Regeneration of the F.O immune sensor using TEA buffer.

Table 2: Immunoassay Techniques Developed for Pesticides Detection*

Pesticide	Reported detection limit	Method	Reference
Fungicides			
Benomyl and metabolites	0.10 ng ml-1 1.25 ng ml-1 350 ng ml^{-1}	FIA RIA ELISA	** ** **
Fenpropimorph	13 pg ml^{-1}	ELISA	**
Metalaxyl	63 pg ml^{-1}	ELISA	**
Triadimefon	1.0 ng ml^{-1}	ELISA	**

Herbicides			
2,4-D	10 ng ml^{-1} 1.0 ng ml-1 500 ng ml-1	ELISA	[32] [33] [34]
2,4-D- 2,4,5-T	0.013, 0.10 ng ml^{-1}	RIA	**
Acifluorfen	1.5 - 4 ng ml^{-1}	ELISA	[35]
Alachlor	0.2 ng ml^{-1}	ELISA	[36]
Atrazine	100 ng ml-1 1.0 ng ml^{-1}	ELISA	**
Bensulfuron	0.03 ng ml^{-1}	ELISA	[37]
Chlorimuron-ethyl	27 ng ml^{-1}	ELISA, IAP	[38]
Chlorsulfuron	0.10 ng ml^{-1}	ELISA	**
Dicamba	2.3 ng ml^{-1}	ELISA	[39]
Diclofop-methyl	23 ng ml^{-1}	ELISA	**
Imazaquin	0.40 ng ml^{-1}	ELISA	[40]
Metolachlor Alachlor Acetochlor	0.06 ng ml^{-1} 0.30 ng ml^{-1} 0.40 ng ml^{-1}	ELISA	[41]
Metolachlor	0.20 - 1.20 ng ml^{-1}	ELISA	[42]
Metosulam	0.30 ng ml^{-1}	ELISA	[43]
Metribuzin	0.50 ng ml^{-1}	ELISA	[44]
Molinate	3.0 ng ml^{-1} 0.40 -1.20 ppb	ELISA	** [45]
Paraquat	10, 0.10 ng ml^{-1} 0.10 ng ml^{-1}	RIA ELISA	** **
Terbutryn	4.80 ng ml^{-1}	ELISA	**
Pesticide	Reported detection limit	Method	Reference
Insecticides			
Aldicarb	300 ng ml^{-1}	EIA	**
Aldrin	0.70 ng ml^{-1}	RIA	**
Azadirachtin	0.50 ng ml^{-1}	ELISA	[46]
Azinophos-methyl	0.08 ng ml^{-1}	ELISA	[47] & [48]
Benzoylphenylurea	0.50 ng ml^{-1}	ELISA	[49]

Bioallethrin ((S)-cyclopentenyl isomer)	0.50 ng ml^{-1}	ELISA	**
Bioresmethrin	50 ng ml^{-1}	ELISA	[50]
Chlordane	5.0 ng ml^{-1}	EIA	**
Chlorpyrifos	0.10 – 0.14 ng ml^{-1}	ELISA	[51]
Deltamethrin	80 ng ml^{-1}	ELISA	[52]
Deltamethrin Cyperrmethrin Cyhalothrin	20 ng ml^{-1}	EIA	[53]
Deltamethrin	2.0 ng ml^{-1}	EIA	[54]
Dieldrin	0.15 ng ml^{-1}	RIA	**
Dieldrin	50 - 80 ng ml^{-1}	ELISA	[55]
Diflubenzuron	3.9 ng ml^{-1}	ELISA	**
Endosulfan	3.0 ng ml^{-1}, 0.20 ng ml^{-1}	EIA ELISA	** [56]
Fenitrothion	0.30 ng ml^{-1}	ELISA	[31]
Fenoxycarb	0.50 ng ml^{-1}	ELISA	[30]
Flucythrinate	200 ng ml^{-1}	ELISA	[57]
Imidacloprid	17.3 ng ml^{-1}	ELISA	[58]
Paraoxon	25 ng ml^{-1}	ELISA	**
Parathion	4.0 ng ml^{-1}	RIA	**
Permethrin	0.50 ng ml^{-1}	ELISA	[59]

*See Reference Section
**[60] *Source: Jung et al., (1989)*

Advantages of ELISA include high volume output and relative low cost compared to GC/MS or HPLC. However, it is slow (each incubation) step takes hours, is laboratory-bound and requires a highly trained technician to operate. Most ELISA assays use immune sera directly. This requires continuous calibration since the polyconal mix of various Abs changes with subsequent collection of sera from the same animal and between animals. mAbs provide a uniform source of Abs, all having a single affinity for the target chemical. The mAbs are produced by fusion, better known as "hybridoma technology", and more recently by cloning the Ab genes in bacteria using recombinant technology. In both cases, the mouse is immunized several times by an Ag, until a high titer of Abs develops. To obtain hybridomas, the spleen is minced and the β lymphocytes are harvested, then fused with immortal myeloma cells. The cell that secretes the desired Ab is isolated and cultured.

mAbs are also produced by recombinant technology, where RNA, extracted from β lymphocytes, is used to generate a cDNA library. The gene fragments of light and heavy chains (that from the binding sites of the Ab) are excised connected by a DNA linker and spliced into a phage expression system. The phage library is screened for the desired Ab and its gene is expressed in *E.coli.* The bacterial culture is easily expanded to yield large quantities of uniform Abs and can survive forever at low cost compared with the mammalian hybridoma cell. The latest developing in recombinant technology is the availability of γ-globulin phage libraries. Such a super-library represents the almost complete repertoire of Ab genes in the animal. We have a human library that contains 10 billion IgG genes. The library is simply screened and the few phages that bind the Ag provide the DNA for an Ab that can be mutated *in vitro* to obtain a higher affinity Ab for this Ag. This biotechnology eliminates the need to immunize animals, speeds Ab production and reduces its cost.

There are also great advances taking place in developing ion selective chemical field effect transistors using Abs as the sensing elements. These biosensors, which are built on silicon chips and packaged in very small space, promise to revolutionize environmental monitoring industrial process control and quality control in the food industry, and different environmental components.

The results presented in this study and in the literature reports suggest that it is usually rewarding and of great benefit to develop several haptens with different spacer arm locations and sizes, various conjugation techniques, and different carrier proteins during ELISA development [29, 30, 31, 32, 33, 34]. Further studies with other assay developed systems and for samples from different environmental components, especially for persistent organic pollutants (POPs), are in active progress.

Acknowledgements

This work was supported by US-Egypt joint US-Egypt Science and Technology Research Program Project MAN 2-004-002-98. We are gratefully acknowledge the helpful comments and advices in this work by Drs. M.E Eldefrawi and A.T. Eldefrawi. In addition, we are indebted to M.E. Eldefrawi, Department of Pharmacology & Expt. Therapeutics, School of Medicine, University of Maryland at Baltimore, where most of this work was performed. We also thank A. Bedair for his technical assistance, especially the computer analysis and work.

References

[1] N. Mansour, Pesticides Production and Application in Egypt, 3rd *International Symposium on Industry & Environment in Developing Countries*, Alexandria (1991) May 29- June 2.

[2] N. Mansour, Strategy, Registration and Use of Pesticides, *Proceedings of the Crop Health Conference*, March 2-24 (1996) 205-226.

[3] N. Mansour, National Industry, Registration and Management of Pesticides in Egypt. "Overview of Current Status, Review of Problems and Needs for the Pesticide Industry, Future Development and Recommendations", *US – Egypt Partnership for Economic Growth and Development. Manufacturing Technologies Workshop*", Dec. 6-9 (1996)

[4] N. Mansour, Pesticides in Aquatic Environment: Problems & Solutions, *Proceedings on Management and Abatement Contamination of Irrigation & Drinking Water, Egyptian Society for Pest Control & Environmental Protection* (1997) 46-52.

[5] D. Pimcntal and C. Edwards, Pesticides and Exosystem, *Bioscience, Int.* N.Y. (1982) 595-600.

[6] M. Aardema and J. MacGregor, Toxicology and Genetic Toxicology in the New Era of "Toxicogenomics": impact of "-omics" Technologies, *Mutat Res* **499** (2002) 13-25.

[7] L.Smith, Key Challenges for Toxicologists in the 21st Century, *Trends Pharmacol Sci* **22** (2001) 281-285.

[8] R. Bandara and S. Kennedy, Toxicoproteomics – a New Preclinical Tool, *Drug Discov Today* **7** (2002) 411-418.
[9] L. Fredrickson *et al.*, Towards Environmental Toxicogenomics – Development of a Flow-Through, High – Density DNA Hybridization Array and its Application to Ecotoxicity Assessment. *Sci Total Environ* **274** (2001) 137-149.
[10] P. Colborn, A Case for Case Study Research, *Am. J. Occup. Ther.*, **50** (1996) 592–594.
[11] R. DeVito, Hospice and Managed Care: The New Frontier, *Am. J. Hosp. Palliat Care* **12** (1995) 2.
[12] E. Den Hond *et al.*, Sexual Maturation in Relation to Polychlorinated Aromatic Hydrocarbons: Sharpe and Skakkeback's Hypothesis Revisited, *Environ Health Prespect* **110** (2002) 771-776.
[13] F. Brady *et al.*, Immunoassay Analysis and Gas Chromatography Confirmation of Atrazine Residues in Water Samples From a Filed Study Conducted in the State of Wisconsin, *J. Agric. Food Chem.* **43** (1995) 268-274.
[14] F. Brady *et al.*, Application of a Triasulfuron Enzyme Immunoassay to the Analysis of Incurred Residues in Soil and Water Samples, *J. Agric. Food Chem.* **43** (1995) 2542-2547.
[15] B. Holmstedt, Pharmacology of Organophosphorus Cholinesterase Inhibitors, *Pharmacol. Rev.* **11** (1959) 567-688.
[16] B. Holmstedt, Structure-Activity Relationships of the Organophosphorus Anticholinesterase Agents In: G. Koelle (ed.), Cholinesterases and Anticholinesterase Agents. Springer - Verlag, Berlin, 1963, 428-485.
[17] B. Ballantyne and T. Marrs, Clinical and Experimental Toxicology of Organophosphates and Carbamates, Butterworth-Heinemann, Oxford, England, 1992.
[18] H. Chambers, Organophosphorus Compounds: An Overview In: J, Chambers and P. Levi, (eds.), Organophosphates – Chemistry, Fate and Effects, Academic Press, San Diego, 1992, 3–17.
[19] L. Sawyer, The Development of Analytical Methods for Pesticide Residues, In: Pesticides Residues in Foods: Technologies for Detection, Office of Technology Assessment, Congress of the United States, OTA -F- 398, U.S. Government Printing Office, Washington, DC, 1988.
[20] J. Manclùs *et al.*, Development of a Chlorpyrifos Immunoassay Using Antibodies Obtained From a Simple Hapten Design, *J. Agric. Food Chem.* **42** (1994) 1257–1260.
[21] J. Manclùs and A. Montoya, Development of an Enzyme-Linked Immunosorbent Assay for 3, 5, 6-Trichloro-2-pyridinol. 2. Assay Optimization and Application to Environmental Water Samples, *J. Agric. Food Chem.*, **44** (1996) 3710-3716.
[22] H. Goodrow *et al.*, Hapten Synthesis, Antibody Development, and Competitive Inhibition Enzyme Immunoassay for S-Triazine Herbicides, *J. Agric. Food Chem.* **38** (1990) 990-996.
[23] J. Langone and H. Van Vunakis, Radioimmunoassay for Dieldrin and Aldrin, *Res. Commun Chem Pathol. Pharmacol.* **10** (1975) 163-171.
[24] E. Coligan *et al.*, Antibody Detection and Preparation In: R Cioco (ed),., Current Protocols in Immunology. John Wiley & Sons, Inc., 1994, 241-271.
[25] A. Larsson *et al.*, Chicken Antibodies: Taking Advantage of Evolution: A Review, *Poulty Science*, **72** (1993) 1807-1812.
[26] T. Glass, Biospecific Binding and Kinetics – a Fundamental Advances in Measurement Technique, Biomedical Products 20 (1995) 122.
[27] K. O' Connell *et al.*, Assessment of an Automated Solid Phase Competitive Fluoroimmuno-Assay for Benzoylecgonine in Untreated Urine, *J. Immuno. Meht.* **225** (1999) 157-169.
[28] N. Nath *et al.*, A Rapid Reusable Fiber Optic Biosensor for Detecting Cocaine Metabolites in Urine, *J. Anal. Toxicol.* **23** (1999) 460-467.
[29] I. Wengatz, *et al.*, Development of an Enzyme-Linked Immunosorbent Assay for the Detection of the Pyrethroid Insecticide Fenpropathrin, *J. Agric. Food Chem.*, **46** (1998) 2211-2221.
[30] J. Sanborn, *et al.*, Hapten Synthesis and Antibody Development for Polychlorinated dibenzo-*p*-dioxin Immunoassays, *J. Agric. Food Chem.*, **46** (1998) 2407-2416.
[31] N. Danilova, ELISA Screening Monoclonal Antibodies to Haptens: Influence of the Chemical Structure of Hapten-Protein Conjugates, *J. Agric. Food Chem.*, **173** (1994) 111-117.
[32] J. Manclùs, et al., Development of Enzyme-Linked Immunosorbent Assays for the Insecticide Chlorpyrifos. 1. Monoclonal Antibody Production and Immunoassay Design, *J. Agric. Food Chem.*, **44** (1996) 4052-4062.
[33] F. Szurdoki, *et al.*, Synthesis of Haptens and Protein Conjugates for the Development of Immunoassays for the Insect Growth Regulator Fenoxycarb, *J. Agric. Food Chem.*, **50** (2002) 29-40.
[34] E. Watanabe, et al., Enzyme-Linked Immunosorbent Assay Based on a Polyclonal Antibody for the Detection of the Insecticide Fenitrothion. Evaluation of Antiserum and Application of the Analysis of Water Samples, *J. Agric. Food Chem.*, **50** (2002) 53-58.
[35] T. Lawruk, *et al.*, Quantification of 2,4-D and Related Chlorophenoxy Herbicides by a Magnetic Particle-based ELISA, *Bull. Environ. Contam.*, **52** (1994) 538 – 545.
[36] J. Fleeker, Two Enzyme Immunoassays to Screen for 2,4- Dichlorophenoxy Acetic Acid in Water, *Assoc. Off. Anal. Chem.*, **70** (1987) 874-878.

[37] J. Kevan, Two Analytical Methods for the Measurement of 2,4-D in Oranges: an ELISA Screening Procedure and a GC-MS Confirmatory Procedure, *Pestic. Sci.*, **50** (1997) 135 – 140.

[38] M. Shiro, *et al.*, Polyclonal and Monoclonal Antibodies for the Specific Detection of the Herbicide Acifluorofen and Related Compounds, *Pestic. Sci.*, **51** (1997) 49 – 55.

[39] P. Feng, *et al.*, Development of an Enzyme-Linked Immuno Sorbent Assay for Alachlor and its Application to the Analysis of Environmental Water Samples, *J. Agric. Food Chem.*, **38** (1990) 159 – 163.

[40] J. Lee, *et al.*, Development of an Immunoassay for the Residues of the Herbicide Bensulfuron-Methyl, *J. Agric. Food Chem.*, **50** (2002) 1791-1803.

[41] C. Sheedy, and J. Hall, Immunoaffinity Purification of Chlorimuron-Ethyl from Soil Extracts Prior to Quantitation by Enzyme-Linked Immuno Sorbent Assay, *J. Agric. Food. Chem.*, **49** (2001) 1151 – 1157.

[42] B. Clegg, *et al.*, Development of an Enzyme-Linked Immuno Sorbent Assay for the Detection of Dicamba, *J. Agric. Food Chem.*, **49** (2001) 2168 – 2174.

[43] R. Wong, and Z. Ahmed, Development of an Enzyme-Linked Immuno Sorbent Assay for Imazaquin Herbicide, *J. Agric. Food Chem.*, **40** (1992) 811 – 816.

[44] P. Casino, *et al.*, Evaluation of Enzyme-Linked Immunoassays for the Determination of Chloroacetanilides in Water and Soils, *Environ. Sci. Technol.*, **35** (2001) 4111 –4119.

[45] T. Lawruk, *et al.*, Determination of Metolachlor in Water and Soil by a Rapid Magnetic – Based ELISA, *J. Agric. Food Chem.*, **41** (1993) 1426 – 1431.

[46] J. Parnell, and C. Hall, Development of an Enzyme-Linked Immuno Sorbent Assay for Detection of Metosulam, *J. Agric. Food Chem.*, **46** (1998) 152 – 156.

[47] D. Watts, *et al.*, Evaluation of an ELISA Kit for Detection of Metribuzin in Stream Water, *Environ. Sci. Technol.*, **31** (1997) 1116 – 1119.

[48] M. Kelley, *et al.*, Chlorsulfuron Determination in Soil Extracts by Enzyme Immunoassay, *J. Agric. Food Chem.*, **33** (1985) 962 – 965.

[49] K. Hemalatha, *et al.*, Determination of Azadirachtin in Agricultural Matrixes and Commercial Formulation by Enzyme-Linked Immuno Sorbent Assay, *J. AOAC Int.*, **84** (2001) 116A.

[50] J. Mercader, and A. Montoya, Development of Monoclonal ELISA for Azinophos-Methyl. 1. Hapten Synthesis and Antibody Production, *J. Agric. Food. Chem.*, **47** (1999) 1276 – 1284.

[51] J. Mercader, and A. Montoya, Development of Monoclonal ELISA for Azinophos-Methyl. 2. Assay Optimization and Water Sample Analysis, *J. Agric. Food. Chem.*, **47** (1999) 1285 – 1293.

[52] S. Wie, and B. Hammock, Comparison of Coating and Immunizing Antigen Structure on the Sensitivity and Specificity of Immunoassays for Benzoyl Phenyl Urea Insecticides, *J. Agric. Food Chem.*, **32** (1984) 1294 – 1301.

[53] A. Hill, *et al.*, Quantitation of Bioresmethrin, a Synthetic Pyrethroid Grain Protectant, by Enzyme Immunoassay, *J. Agric. Food Chem.*, **41** (1993) 2011 – 2018.

[54] J. Manclús and A. Montoya, Development of an Enzyme-Linked Immunosorbent Assay for the Insecticide Chlorpyrifos. 2. Assay Optimization and Application to Environmental Waters, *J. Agric. Food Chem.*, **44** (1996) 4063-4070.

[55] A. Queffelec, *et al.*, Hapten Synthesis for a Monoclonal Antibody Based ELISA for Deltamethrin, *J. Agric. Food Chem.*, **46** (1998) 1670 – 1676.

[56] N. Lee, *et al.*, Development of Immunoassays for Type II Synthetic Pyrethroids. 2. Assay Specificity and Application to Water, Soil, and Grain, *J. Agric. Food. Chem.*, **46** (1998) 535 – 546.

[57] N. Lee, *et al.*, Development of Immunoassays for Type II Synthetic Pyrethroids. 1. Hapten Design and Application to Heterologous and Homologous Assays, *J. Agric. Food. Chem.*, **46** (1998) 520 – 534.

[58] N. Kawar, *et al.*, Comparison of Gas Chromatography and Immunoassay Methods in Measuring the Distribution of Dieldrin in Rainbow Trout Tissues, *J. Environ. Sci. Health. B.*, **36** (2001) 765 – 774.

[59] N. Lee, *et al.*, Hapten Synthesis and Development of ELISA for Detection of Endosulfan in Water and Soil, *J. Agric. Food. Chem.*, **43** (1995) 1730 – 1739.

[60] M. Nakata, *et al.*, A Monoclonal Antibody-Based ELISA for the Analysis of the Insecticide Flucythrinate in Environmental and Crop Samples, *Pest. Manag. Sci.*, **57** (2001) 269 –277.

[61] N. Lee, *et al.*, Development of an ELISA for the Detection of the Residues of the Insecticide Imidacloprid in agricultural and Environmental Samples, *J. Agric. Food Chem.*, **49** (2001) 2159 –2167.

[62] J. Skerritt, *et al.*, Analysis of the Synthetic Pyrethroids, Permethrin and 1 (R)-Phenothrin, in Grain Using a Monoclonal Antibody-Based Test, *J. Agric. Food. Chem.*, **40** (1992) 1287 – 1292.

[63] F. Jung, *et al.*, Use of Immunochemical Techniques for the Analysis of Pesticides, *Pestic. Sc.*, **26** (1989) 305 – 317.

Prospects for Holographic Optical Tweezers

Joseph S. PLEWA, Timothy DEL SOL, Robert W. LANCELOT, Ward A. LOPES, Daniel M. MUETH, Kenneth F. BRADLEY, Lewis S. GRUBER
Arryx, Inc.
316 N. Michigan Ave. Suite CL-20
Chicago, IL 60601

1. Introduction

Since their introduction in 1986 [1], so-called optical tweezers or optical traps have permeated research in the biological and physical sciences. In biology, the explosion of research in cellular and molecular biology has demanded a method for controlling cells, cellular organelles, and biopolymers, making the demand for tweezers grow [2-7]. In physics, the study of phenomena at the microscopic scale has become an important area of research. Such phenomena provide an accessible testing ground for ideas about matter at the atomic scale [8], and are interesting in their own right as complex systems. The need to control and perturb these systems has driven optical tweezer research [9-12].

Ashkin's original work provided scientists with a "hand" with which to grasp objects in the microscopic realm. As experimental complexity grew, the need for multiple hands for manipulating several experimental components became clear. The impetus for holographic optical trapping grew as a natural result of this emerging necessity [13-15]. Whereas single optical traps involve simply sending a single laser beam through the objective lens of a microscope, holographic optical traps first bounce the beam off of a spatial light modulator. The spatial light modulator is a liquid crystal display designed to modulate the phase of the laser front. The modulator thereby acts to introduce a complex diffraction pattern to the laser light distribution, i.e. it produces a hologram. This hologram then enters the objective lens to be focused to a diffraction-limited spot, just as in the case of the single tweezer.

There are three major advantages to holographic optical trapping over non-holographic trapping methods. The first is the number of traps. The BioRyx™ 200 system can create up to 200 individual optical traps, each independently controllable. Because there is no time-sharing of the laser beam, as there is with traps generated by scanned mirrors, there are fewer restrictions on the number of traps, the laser power required, and the spatial separation of the traps. The second advantage is three-dimensional control. Each of the individual traps can be placed in three-dimensions independently. Before holographic optical trapping, the only way to modify the three-dimensional position of the traps was to move the microscope stage, necessarily moving the entire of field-of-view at the same time. The third major advantage of holographic optical trapping is the ability to superimpose phase profiles on individual traps. This allows for the creation of exotic modes of light, like optical vortices and optical bottles, which provide added trapping versatility. Optical trapping takes advantage of the fact that the trapped particle has a higher index of refraction than the surrounding medium. Traditionally, particles which reflect, absorb, or have an index of refraction lower than that of the surrounding medium have not been trapped. However, an optical bottle, which is essentially a tweezer with a dark region surrounded by higher intensity regions, can trap all of these particles which are not trapped with normal tweezers. Optical vortices enhance trapping by imparting angular momentum to particles, causing the particles to spin on their axes [16].

2. Optical Tweezers in Microsphere-Based Systems

Since Ashkin's original paper [1], a great deal of the work in optical trapping has been performed on colloidal microspheres. Monodisperse microspheres made from a large assortment of materials are commercially available in any size accessible to traps, making the spheres ideal candidates for trapping experiments. Furthermore, theoretical treatments of optical trapping of microspheres are tractable because of the sphere's shape and uniform index of refraction [17]. This is useful not only because it facilitated early understanding and development of trapping, but also because it allows tweezers to be employed as force transducers with a straightforward calibration procedure [18].

Advances in biochemistry and the surface chemistry of colloidal particles have made possible a large range of affinity studies using optical tweezers as force transducers. In a typical study, microspheres coated with some antigen are pulled off of a surface coated with the corresponding antibody. Displacement of the spheres in the calibrated tweezer indicates the force required to break the bond. Systems studied to-date include bovine serum albumin (BSA) and anti-BSA [19] and *Staphylococcus* protein A and immunoglobulin G [20].

The same advances that have made affinity studies possible, have also enabled single-molecule studies. In a typical experiment, a single molecule of some biopolymer is attached between two colloidal spheres or a sphere and a surface, and the molecule is stretched, yielding a force versus extension curve. The canonical molecule for this type of study has been DNA [21,22], which continues to be studied with increasingly adroit tweezer manipulations under increasingly complex enzymatic conditions [23-30]. Recently, a large assortment of molecules of biological interest has been studied through optical manipulation [31-33]. For example, collagen [34,35], RNA [36], actin/myosin [37], kinesin [38], and microtubules [39,40] have all been manipulated by optical traps in order to gain structural or functional information.

Using a microscope, it is possible to study material properties of fluids or solids on length scales much shorter than those typical of bulk measurements. To perform such a study, some manner of perturbation must be imposed on the matter, and the material response must be measured. Optical tweezing of microspheres that have been attached to a material of interest has made such studies possible. For example, both cellular membranes [41-43] and the extra-cellular matrix [44] have been studied in such a fashion.

Optical trapping has also been employed as a construction technique. Traps have been used to assemble and polymerize colloidal structures into microfluidic devices [45]. Because colloidal spheres are of a size that is useful for photonic band-gap materials, much effort has been directed towards using tweezers to arrange or template growth for photonic crystals. Traps have also been used as a tool for manipulating hazardous objects [46, 47].

The development of holographic optical trapping has significantly extended the utility of optical tweezers. For example, affinity studies may now be performed massively in parallel, increasing throughput and improving statistics. Large arrays of beads, coated with antibodies may be trapped (Figure 1) and probe beads with a variety of antigens coated on the surface may be pulled from each particle in turn (Figure 2). Furthermore, the spatial light modulator that is employed for generation of the holograms, inherently involves a computer interface for control of the traps. This means that a computer-controlled system like the BioRyx™ 200 system employs software control of the motion of the tweezers, allowing for the possibility of completely automated measurements. The capability of automation, combined with the extremely precise control of holographic traps suggests the possibility for elaborate studies.

Construction of microscopic devices using microspheres provides another example of the advantages of holographic optical trapping [48]. In addition to the potential for higher-speed construction, and automated construction, holographic trapping has the significant capability of manipulating traps independently in three dimensions (Figure 3). This means that non-planar microstructures may be easily created. A classic problem in using spheres subject to Brownian motion for construction is the difficulty of having enough spheres to build useful devices, but not so many that there are no open regions to serve as a construction space. Dynamic holographic traps circumvent this problem by allowing the user to sweep spheres to a designated perimeter, where they rest until needed as building blocks (Figure 4). Dynamic holographic tweezing also has certain advantages that emerge from the coordinated movement of a large number of traps. For example, assembly of high-quality colloidal crystals is difficult because of the inevitable appearance of defects. Sweeping holographic arrays of traps across a defect-ridden crystal provides an annealing method that removes the defects [49].

Holographic tweezers can also serve a critical purpose in driving microfluidic components [50, 51]. Lab-on-a-chip applications, which allow sophisticated studies with microscopic quantities of reagents, require microfluidic chambers with components like pumps and valves. Figure 5 shows a glass shard about five microns in diameter undergoing controlled rotation by a laser tweezer. A rotating shard about this size would make a convenient pump when inserted into a microfluidic channel.

Hazardous particle handling provides yet another example of holographic optical trapping in microsphere-based systems. Multiple, dynamic, holographic tweezers allow control of many hazardous particles in three-dimensions. Furthermore, rotation of the particles is possible, using optical vortex modes which are standard in the BioRyx™ 200 system. Because of the large number of tweezers available, backup tweezers can be generated which will trap particles should some exterior forces cause them to drop. Finally, multiple tweezers can grab the same particle, which decreases the likelihood of unwanted effects like heating or ablation.

3. Optical Tweezers in Cell-Based Systems

Ashkin demonstrated the successful tweezing of viruses and bacteria in the year following his demonstration of the apparatus on dielectric spheres [52, 53]. Since then, there has been a proliferation of applications that involve tweezing objects of biological origin [54, 55]. In addition to prokaryotes and viruses, a large variety of protists such as *Tetrahymena thermophila* has been successfully tweezed. Furthermore, both somatic cells such as erythrocytes and epithelial cheek cells, and germ line cells such as spermatozoa [56] have been trapped and manipulated. Because cellular damage from irradiation by highly-focused laser light is a serious threat, researchers have sought indirect methods for manipulating cells, such as tagging the cells with diamond micro-particles and then tweezing the diamond particles [57]. Cell manipulations have included cell orientation for microscopic analysis [58] as well as stretching cells [59]. Tissue cells have also been arranged with tweezers in vitro in the same spatial distribution as in vivo [60, 61].

In addition to the cells themselves, optical tweezers have been used to manipulate cellular organelles [62], such as vesicles transported along microtubules [63], chromosomes [64], or globular DNA [65-66]. Ponelies uses optical tweezers to simulate microgravity in algae by lifting gravity-sensitive organelles [64]. Objects have also been inserted into cells [67].

A variety of sorting processes for biological purposes is also possible with optical tweezers [68]. Cell sorting for assays and chromosome collection and sorting to create libraries [64, 69] have already been demonstrated. Cell assays for drug screening have also been developed [70].

Holographic optical trapping promises large benefits for manipulation of biological organisms. Cell sorting applications, for example, may be made faster by handling a larger number of cells at a time. Figure 7 shows a microfluidic system which is sorting objects based on size, by utilizing an array of tweezers which sweeps by the interaction volume and preferentially grabs large objects. More complex sorting criteria may also be implemented. For example, membrane elasticity has been shown to be an indicator of cellular health. A cell may be flowed into a chamber, the cellular membrane plucked with a few tweezers, and the cell moved into a particular outlet channel based on the response of the membrane to the perturbation. Furthermore, even static arrays of traps have been shown to deflect particles in a flow in a manner which shows great promise for sorting applications [71].

Control of multiple cells like those in tissue is augmented by the freedom to move several cells at once and arrange them in three dimensions. This clearly aids in efforts to arrange cells *in vitro* as closely as possible to their structure *in vivo*. It has been shown that the growth of cells can be influenced by light [72], which means that in addition to the brute force arrangement of tissue cells, they may also be coaxed into growing in useful patterns by lower intensity arrays of tweezers.

Manipulation of cells in general, is made safer by the availability of multiple beams. Like a bed of nails, multiple tweezers ensure that less power is introduced at any particular spot in the cell. This eliminates hot spots and reduces the risk of damage. Any destructive two-photon processes benefit greatly since the absorption is proportional to the square of the laser power. Just adding a second tweezer decreases two-photon absorption in a particular spot by a factor of four. Figure 8 shows a *Tetrahymena thermophila* cell held in place by an array of tweezers. Large cells like *Tetrahymena* require a large amount of laser power for effective trapping. Putting the required power into a single trap causes immediate damage to the cell.

Finally, manipulation of even just a single cell is greatly enhanced by utilizing holographic optical trapping. Figure 9 shows a sequence of images of a single epithelial cheek cell. The cell is manipulated by a line of tweezers, which lift the cell along the perimeter on one side. The resulting rotation allows a 360 degree view of the cell. In addition to the advantage for viewing of biological samples, there also exists the ability to orient samples stably, which has clear benefit for studies such as scattering experiments which have a strong dependence on orientation of the sample.

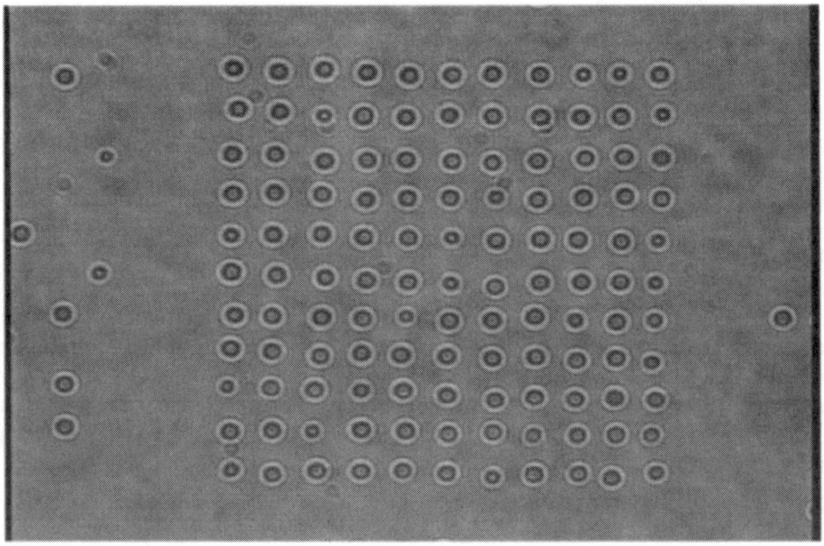

Figure 1. An array of 121 2.25 micron silica spheres. The ability to assemble and maintain such an array suggests the capability for doing micro-construction as well as massively parallel affinity studies and cellular assays.

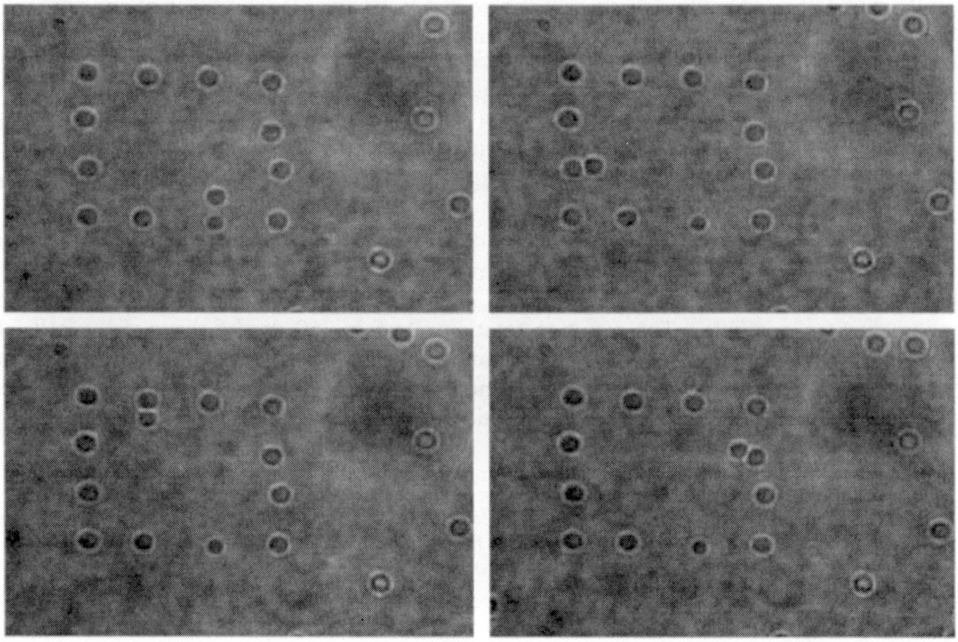

Figure 2. A sequence of images of 2.25 micron silica spheres, indicating the ability for parallel, automated affinity studies. Note the single sphere inside the 4-by-4 square perimeter that moves around and interacts with the other spheres.

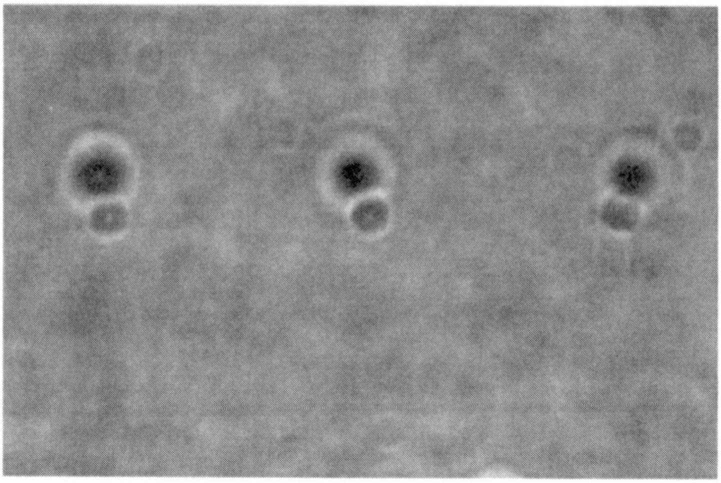

Figure 3. Three-dimensional positioning of 2.25 micron silica spheres. Three of the spheres lie about 5 microns below the focal plane of the microscope objective lens, while three lie about 10 microns above the focal plane.

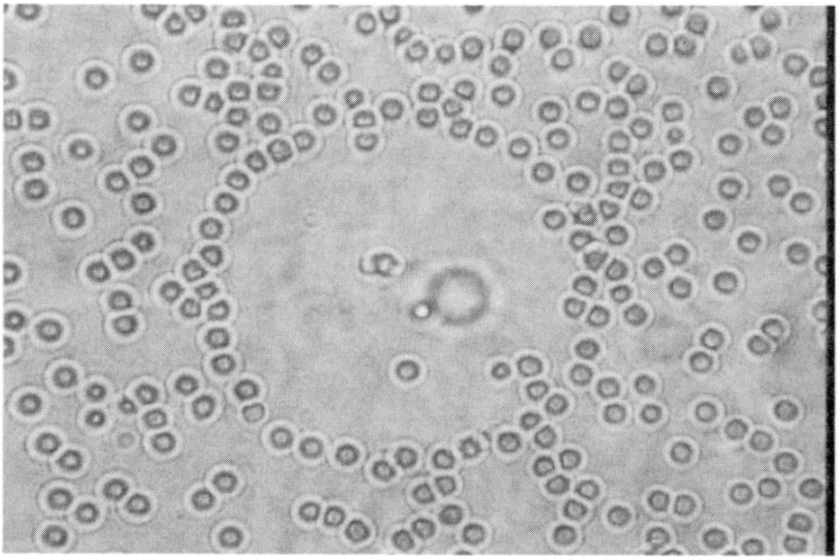

Figure 4. An optical fence generated by sweeping a radially-moving array of traps outward from the center of the screen. The dynamic trapping pattern can be re-run at the user's discretion to clear out straggling particles. This technique establishes a working area for construction or cellular studies, yet maintains a large supply of particles available on the perimeter of the fence. The particles are 2.25 micron silica spheres.

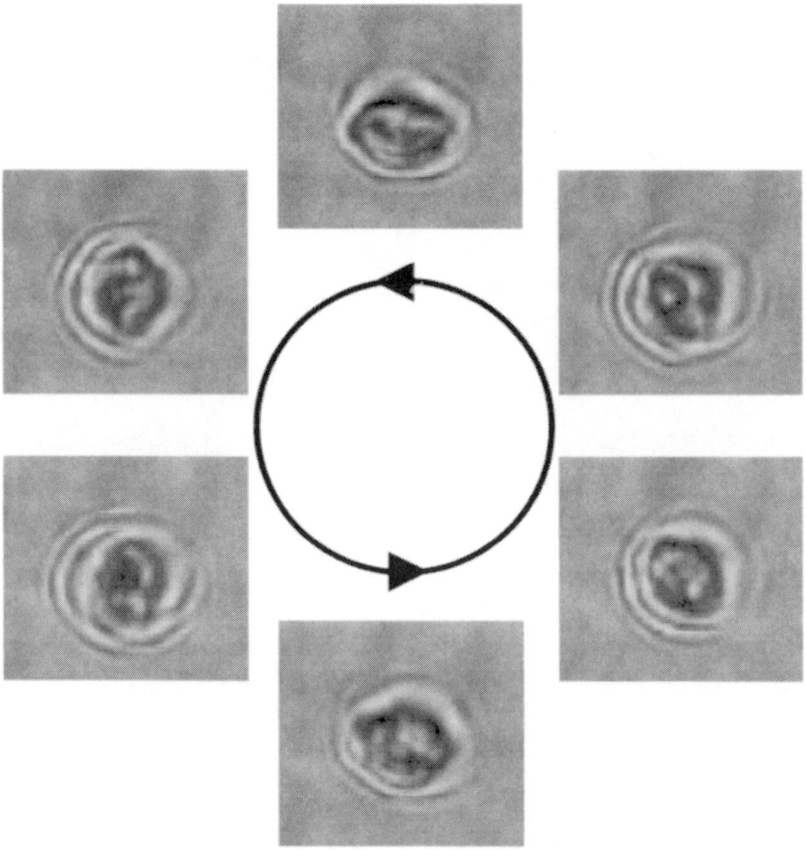

Figure 5. A sequence of images indicating the optical rotation of a glass particle. The particle is about 5 microns in diameter. Control over rotation speed was achieved by controlling the laser power.

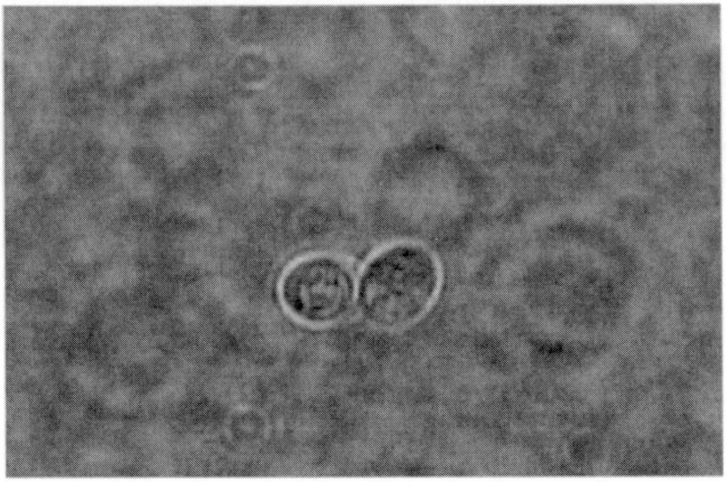

Figure 6. Yeast cells (*Saccharomyces cerevisiae*). Two cells are brought into contact with each other and then separated, demonstrating the potential for affinity studies.

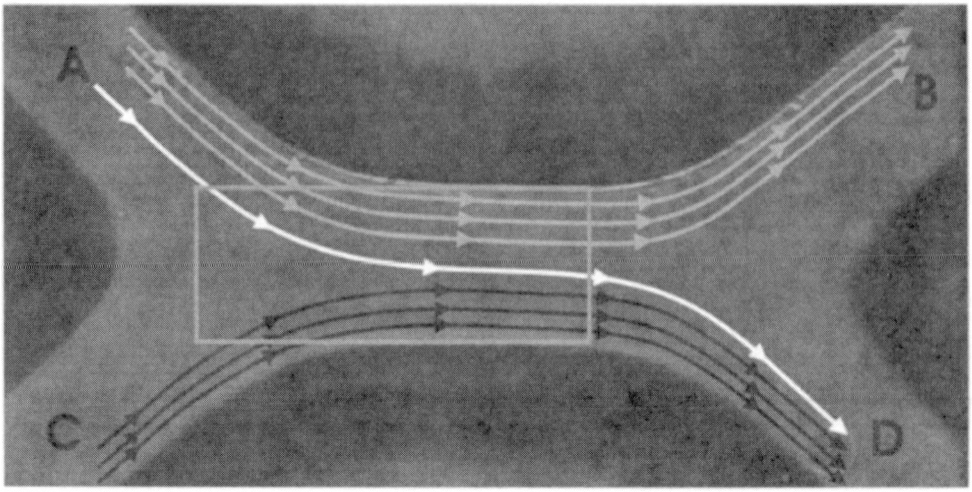

Figure 7. A particle sorting chamber. The sorting system relies on an array of optical traps in the central region, which sweep continually down, dragging larger particles (> 4.5 microns) into channel D, but leaving smaller particles in channel B. The inlet A has a mixed stream of input particles, and the inlet C carries just water. The microfluidic chamber carries a laminar flow, thereby requiring just a small nudge to get from the AB flow to the CD flow.

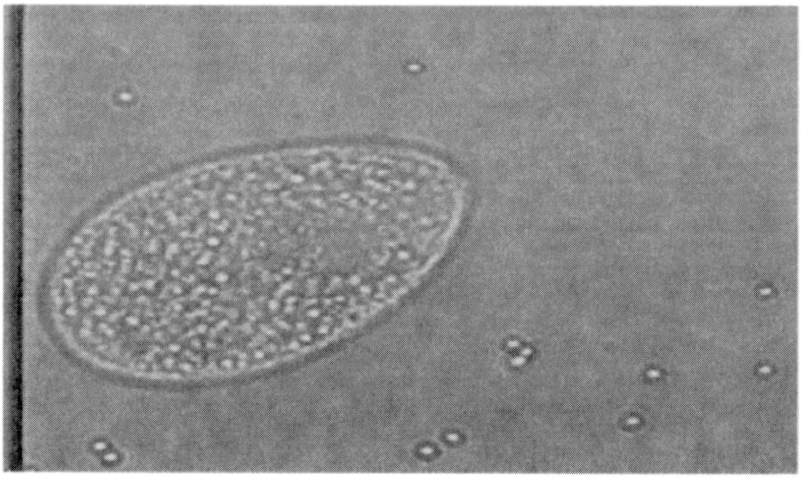

Figure 8. *Tetrahymena thermophila* cell controlled by a net of optical traps. The trapping net approach reduces the power required in each laser spot, thereby minimizing the likelihood of damage to the specimen.

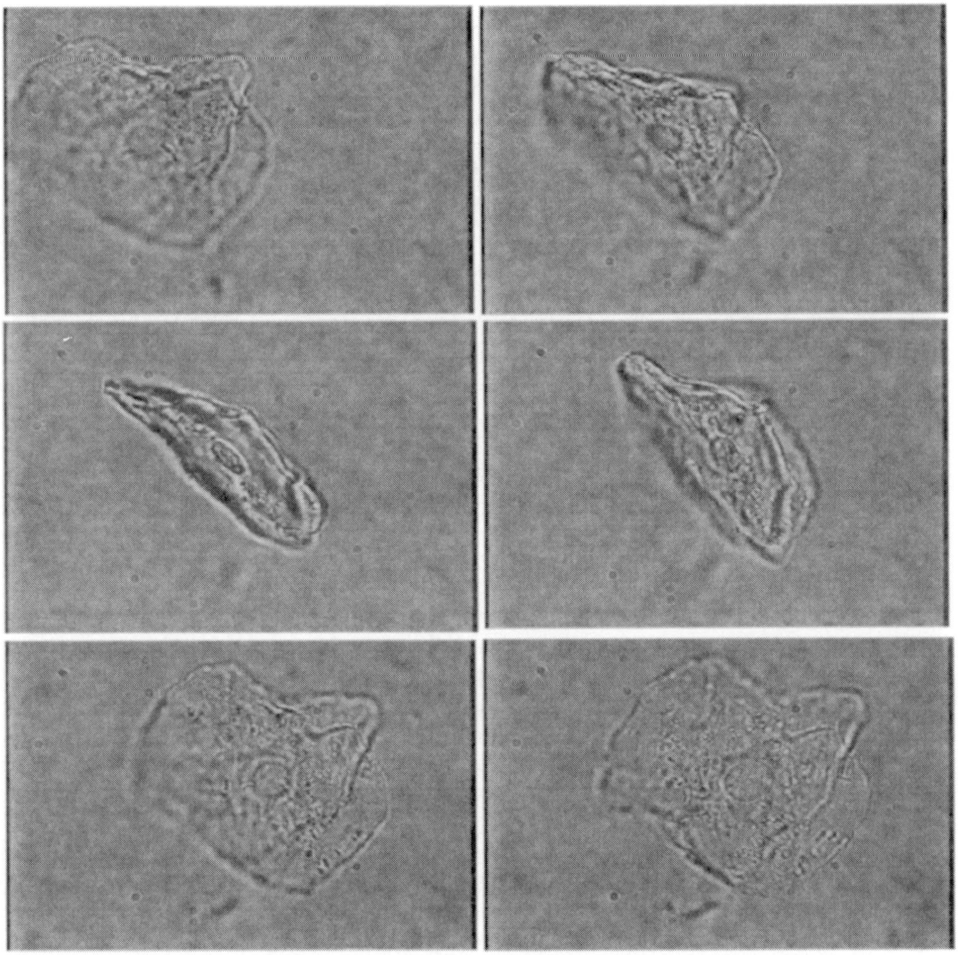

Figure 9. Sequence of images showing the rotation by 180 degrees of an epithelial cheek cell by a line of optical tweezers which grasp the cell by one edge.

4. Non-standard Applications of Optical Tweezers

In addition to the panoply of optical tweezer applications involving cellular systems and microsphere-based systems, there exists a wide range of applications that do not inherently involve tweezing colloidal microspheres, or cells and their various constituents. These applications are generally of two types. There are those that involve trapping, but the objects trapped are of a different type or size than those discussed above. There are also applications that do not involve trapping at all, but rather use the optical tweezers as a source of light, or heat, for example, with fine resolution.

Because laser tweezers inherently tend to aggregate particles with high dielectric constant, even when the particles are considerably smaller than the wavelength of light, any solution of such particles will show density fluctuations under tweezer illumination. This effect has been exploited to nucleate crystals around tweezer focal points in super-saturated solutions [73]. Different polarizations of laser light have been shown to result in the crystallization of different polymorphs of the amino acid glycine [74]. This suggests the possibility of using more exotic forms of tweezer light to form other polymorphs, or to control crystallization in other systems. In addition to nucleating supersaturated solutions, tweezers have also been used to select and move single seeds in solution [75, 76].

In addition to tweezing solid particles, it has also proven possible to tweeze micro-droplets of liquid or gas suspended in another fluid. For example, droplets of water and ethylene glycol suspended in liquid paraffin have been tweezed. The technique has also been utilized for obtaining the Raman spectra of picoliter quantities of *p*-cresol [77] and of stratospheric aerosol particles. The ability to tweeze liquid particles has also been used to create phase separation in binary liquids [78].

Optical tweezers have also made possible a new form of microrheology. Viscosity of liquids may now be measured for femtoliter quantities of liquid. The Brownian motion of trapped particles is studied, yielding the local viscosity [79, 80]. The technique has been used to measure viscosity changes resulting from release of biopolymers into the fluid surrounding a cell.

The high intensity of laser light at the focal point of the tweezer has led to the development of applications employing the tweezers for laser-writing. For example, tweezers can be used to excite two-photon processes at their focal point, allowing photopolymerization. Galajda use tweezers to not only fabricate micro-structures, such as rotors, but then also to drive the components. [50, 51]. Another form of laser-writing has also has been demonstrated, in which optical tweezers are used as a nozzle to shoot particles out onto some surface to which the particles adhere [81, 82]. For these applications, the laser is intentionally focused to a point with a weak axial gradient, so that radiation pressure dominates axial trapping, resulting in the controlled ejection of particles.

Although optical tweezers are usually used to manipulate particles with a characteristic size of about a micron, techniques for controlling much smaller particles have been developed. Single fluorescent molecules have been manipulated by using the tweezer to excite surface plasmon resonance on a metal tip. Nanotubes and nanorods have also been trapped and forced to rotate [83, 84].

Katsura et al. have employed optical tweezers to develop a microreactor for studying chemical reactions [85]. The application has many of the features of the lab-on-a-chip application described above, but doesn't involve the underlying platform. Instead, minute quantities of reagents are placed in water droplets surrounded by oil. The droplets are fused to start a reaction.

Crystal-seeding provides a final example of the versatility of holographic optical trapping. Creating multiple nucleation sites, spaced in three-dimensions, speeds crystal growth, and provides countless possibilities for studying the formation of domains and topological defects in crystals. Because polarization affects the crystallization procedure, it is realistic that the various modes of light that may be generated holographically serve to select between various possible crystalline states.

5. Application of Holographic Optical Tweezers to the Construction of Biological Sensors

Optical steering technology may be used to direct microscopic laser beams to perform manufacturing, processing and analysis. Such systems have applications for the manipulation of hazardous materials inside sealed containers and for manufacturing our sensor/detector products.

A nanosensor platform for use in sensing and detecting biological, chemical and radiological threats in a wide variety of types of samples such as water, air, food and soil may be constructed using the BioRyx™ 200 system available from Arryx, Inc.. The nanosensor platform includes a fiber optic bundle that extends outward from a handheld device. Each sensor for the platform is positioned at the end of an optical fiber, and each sensor fiber may deploy 100 tests, so that 10,000 separate tests may be configured in an area the size of a quarter

When present, substances targeted by the test change the relative positions of test-carrying particles gelled in the matrix of the sensor element. The properties of light in the optical fiber are altered by the change in the matrix and are rapidly transmitted as an optical signal. The sensor elements may be disposable, and may be easily replaced when contaminated or to select different tests. Tests may be adapted from those available from the National Center for Food Safety and Technology, USAMERID, or the Centers for Disease Control.

The reader may be a COTS handheld device and can be configured to transmit data to central locations via wireless links. The device displays simultaneous real-time analyses of all tests for the operator as well.

The optical signals from the device are by their nature not susceptible to contamination or interference. The equipment is thus well suited to hazardous environments and to remote deployment and operation as by being dropped out of aircraft.

An operator may examine a large area by sweeping the sensor bundle as in a brush where each fiber in the brush is a sensor element being swept across surfaces or in an air or water flow. Teams of operators may sweep assigned portions of a contaminated area. Results are reported simultaneously, and correlated in the field to develop pattern and intensity analyses.

6. Overview of the Bioryx™ 200 System Software and Capabilities

Until the development of Holographic Optical Tweezers (HOT) technology by Arryx, Inc., optical trapping systems available were limited in their utility. These early systems were costly and were only able to produce as many as 1-8 traps, which in many cases had fixed trap placement and were of low quality in trapping abilities.

With HOT technology, the limitations of earlier optical trapping systems are removed. The number of traps produced with HOT can range on the order from a few hundred to thousands. The number of traps is highly scalable. Each trap produced with HOT may be independently placed within a three dimensional space and can be modeled to have a variety trapping properties.

To make HOT technology accessible to researchers and system integrators, Arryx, Inc. provides three levels of software access. At the highest level is the application software BioRyx™ 200 system. The next level is the Arryx's Application Programmers Interface (API) for Windows™ 2000. This is available for developers who prefer to develop their own optical trapping application or have an existing application and would like to adapt it for HOT capabilities. Then there is the Arryx's HOT Library. This is intended for application were tight software integration is required, such as in embedded systems.

This document gives a high level overview of the different levels of software access to Arryx's Inc. HOT technology.

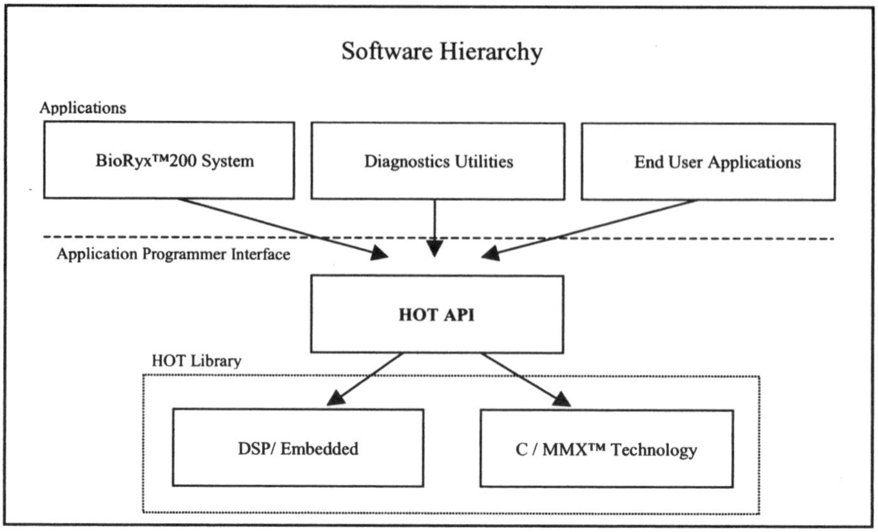

Figure 10. Software Hierarchy

GUI

To drive HOT technology Arryx, Inc. created intuitive Graphical User Interface (GUI) programs which run under Windows™ 2000. The GUI programs are meant to give users easy access to optical trapping without having to write any code.

BioRyx™ 200 System Software

The BioRyx™ 200 system is the first commercially available optical trapping system utilizing HOT technology. The BioRyx™ 200 system software is used to manage the placement and operation of as many as 200 traps. Traps are created under a Nikon TE-2000 microscope. A video camera is attached to the microscope for video capture into the BioRyx™ 200 system software.

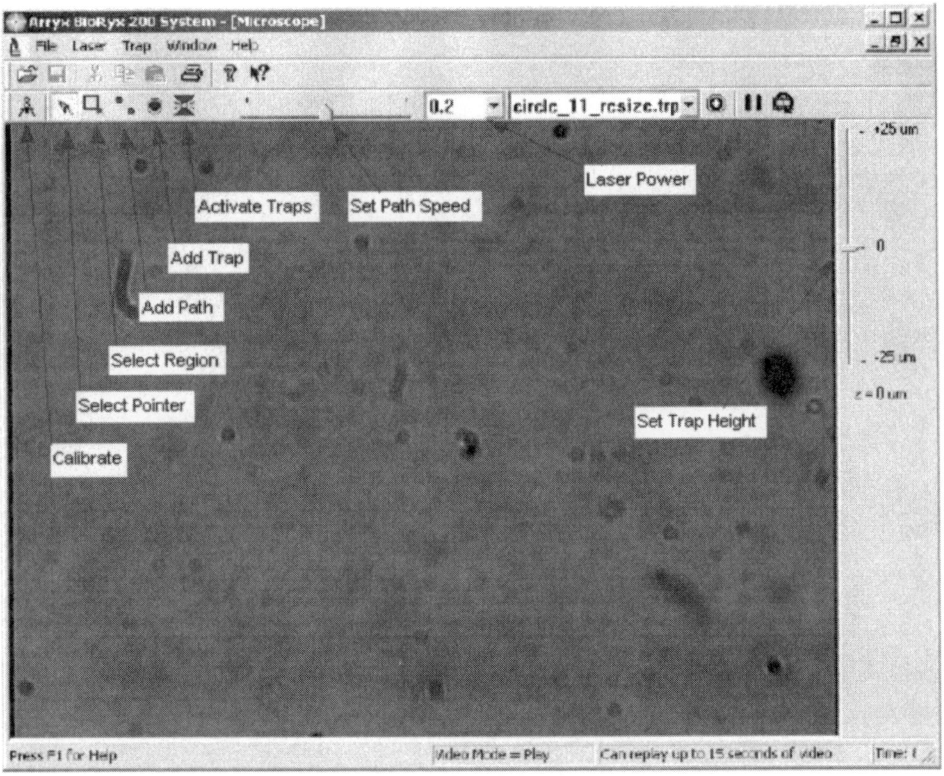

Figure 11. BioRyx™ 200 System Software

Creating Traps

To create a trap, the Add Trap tool is selected from the Microscope window toolbar. The mouse cursor then changes to a crosshair when over the video area of the Microscope window. Placing the crosshair over the location in the Microscope window where the trap is to be placed, then pressing the left mouse button to add the trap icon positions the trap.

To create the trap, the Activate Traps button is pressed in the Microscope window toolbar.

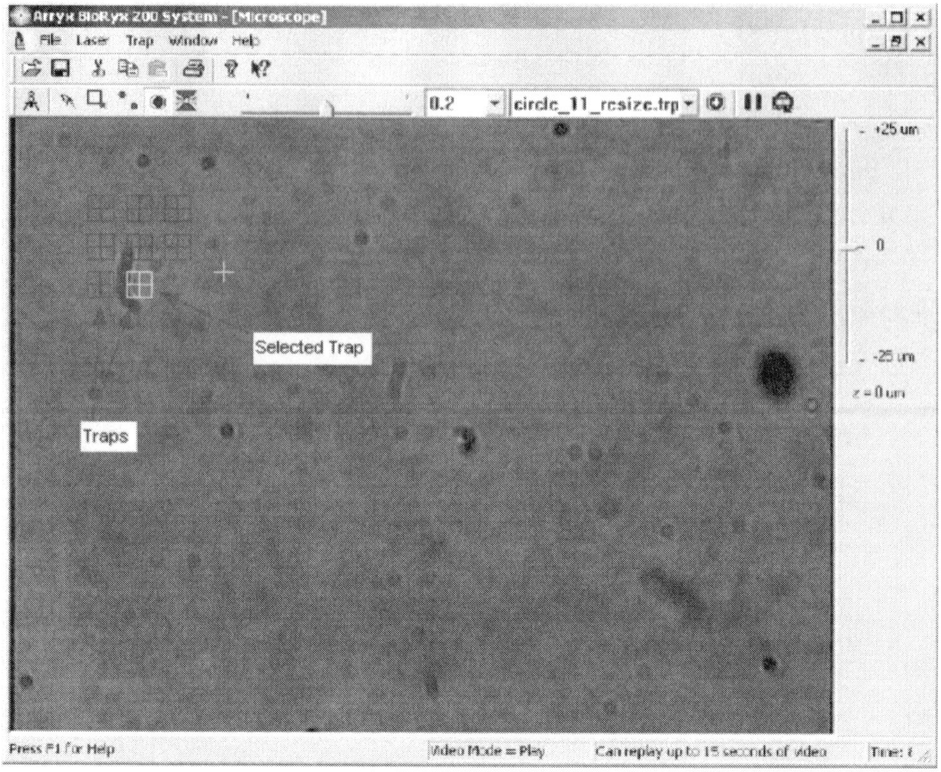

Figure 12. Creating Traps

Creating Trap Paths

To create a path for a trap, the Trap Path tool is selected from the Microscope window toolbar. The mouse cursor changes to the Trap Path cursor when over the video area of the Microscope window. By placing the Trap Path cursor over an existing trap and pressing the left mouse button, a new trap appears with a line segment connecting the two traps. The arrow at the end of the line segment denotes the direction the trap will follow. While the left mouse button is pressed, the new Trap Path may be moved to the desired end point location. Multiple Trap Paths may be added to a trap.

After all the paths for a trap have been created, pressing the Activate Traps button in the Microscope window toolbar starts the trap tracing the created paths.

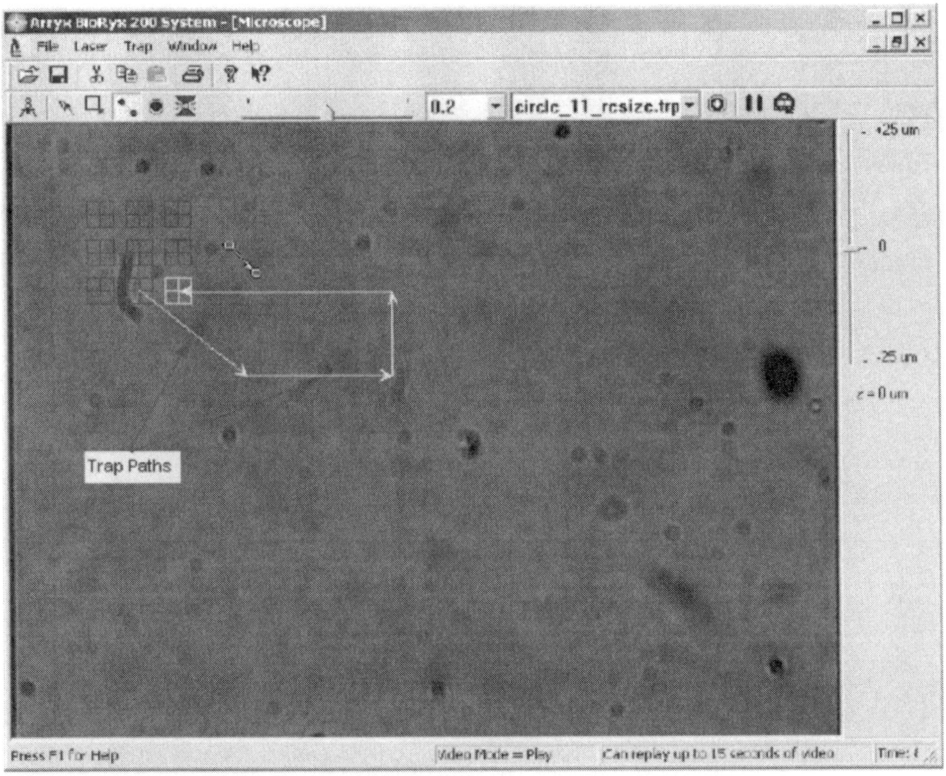

Figure 13. Creating Trap Paths

Trap Speed

After creating a trap with a Trap Path, the speed at which the trap is to follow the path may be adjusted by moving the Speed slider in the Microscope window toolbar. The trap speed may vary from 0.250 µm per second to 1.5 µm per second. A field in the status bar in the lower right corner of the Microscope window shows the current path speed.

3D Traps

Each trap may have an independent vertical location. To set the vertical location of a trap, the mouse cursor is used in the Select mode. To put the mouse cursor in Select mode, the Select Tool is selected from the Microscope window toolbar. The mouse cursor changes to a Pointer when over the video area of the Microscope window.

When selecting an existing trap from the Microscope window with the Pointer, a selected trap will turn green when selected. On the right side of the Microscope window, a vertical slider indicates the current vertical location relative to the focal plane. The vertical distance may range, for example, from 25µm to -25µm. A vertical distance of zero indicates the trap is in the focal plane. A positive vertical distance indicates a vertical location towards the microscope's objective lens. The trap icon appears larger to indicate trap is closer to the objective lens. A negative vertical distance indicates a location away from the microscope's objective lens. The trap icon appears smaller to indicate a trap is further away from the objective lens.

After setting the vertical distance for a trap, pressing the Activate Traps button in the Microscope window toolbar activates the trap with new vertical location.

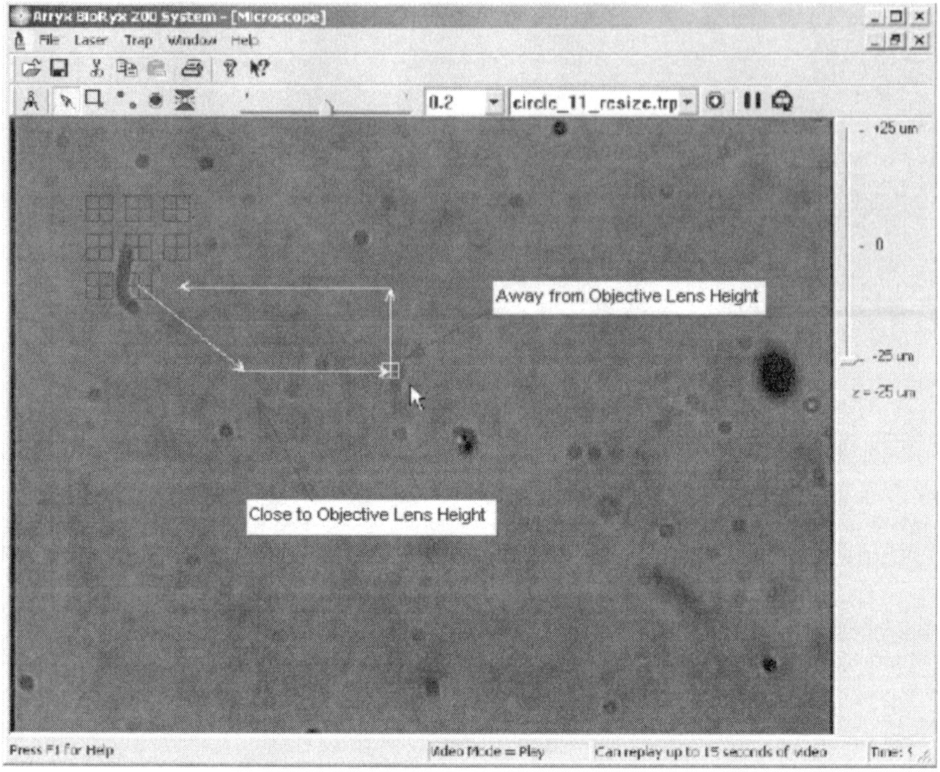

Figure 14. 3D Traps

Vortices

By default all traps created are Point traps. To create a Vortex trap, a user selects and right mouse clicks on a trap icon in the Microscope window. The Trap Properties dialog appears. In the Trap Properties dialog the edit field labeled "Charge" may be changed from -75 to 75. The magnitude of the charge indicates the size of the Vortex and the 'sign' of the charge indicates the direction. A charge of zero creates a Point trap.

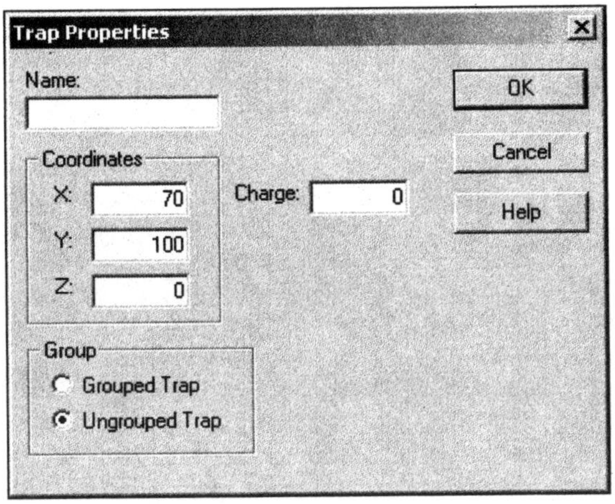

Figure 15. Trap Properties

Laser Power

The laser power maybe used to increase HOT trapping strength. The List Box in the Microscope window toolbar may be used to adjust the laser power from 0.2 to 2.0 watts.

Depending on the sample being trapped, laser power may be used sparingly. Too much laser power may damage the sample.

HOT Application Programmers Interface

Some applications may not require the use of the BioRyx™ 200 system Graphical User Interface (GUI) software, and Researchers and Integrators may have an existing application they would like to adapt to have optical trapping capability. Access to HOT technology may be made available through an API under the Windows 2000® platform. The API interface is accessible from any programming language running under Windows 2000® that is capable of calling functions in a Dynamic Link Library (DLL).

7. Conclusion

Over the last decade, optical trapping has provided critical functionality to research in biological and physical sciences. Holographic optical trapping significantly extends that functionality, facilitating a large number of existing applications for optical tweezers, and opening whole new arenas for research and manufacturing.

Despite its many applications, accessibility to optical trapping has been limited. The BioRyx™ 200 system independently holds, moves, rotates, joins, separates, stretches and otherwise manipulates hundreds of microscopic objects using only laser beams. The ability to perform all of these functions gives rise to multiple applications which have important roles in cancer research, drug development and other therapeutic research, genetic and cell reproduction research and many other areas.

The BioRyx™ 200 system may be used to create dynamically configurable, three-dimensional arrays for proteomics, toxicogenomics and diagnostics assays. The BioRyx™ 200 system's three-dimensional arrays enable greatly enhanced high-throughput screening for accelerating drug discovery and development.

The BioRyx™ 200 system may be used, for example, to collect specified types of cells from a mixed suspension, manipulate cells for enhanced viewing, measure cell-cell interactions and cell-object interactions (e.g. study cell-antigen interactions or bring natural killer cells into contact with a cancer cell), extract nuclear material from specific cells (e.g. chromosomes or DNA) and hold this sample material for further investigation or isolation. BioRyx™ 200 system bioengineering applications include introducing DNA into cells to obtain transgenics and cells for pharmaceutical production. Other applications include arraying cells for screening and seeding cells into a matrix to make artificial tissues.

Trade Marks

Windows™ 2000 is a registered trademark of Microsoft
MMX™ Technology is registered trademark of Intel
BioRyx and Arryx are trademarks of Arryx, Inc.

References

[1] A. Ashkin, J.M. Dziedzic, J.E. Bjorkholm, and S. Chu, Observation of a single-beam gradient force optical trap for dielectric particles, *Optics Letters* **11**(5) (1986) 288.
[2] K. Schutze, I. Becker, K.F. Becker, S. Thalhammer, R. Stark, W.M. Heckl, M. Bohm, and H. Posl, Cut out or poke in – The key to the world of single genes: Laser micromanipulation as a valuable tool on the look-out for the origin of disease, *Genetic Analysis-Biomolecular Engineering* **14**(1) (1997) 1.
[3] K. Schutze, H. Posl, and G. Lahr, Laser micromanipulation systems as universal tools in cellular and molecular biology and in medicine, *Cellular and Molecular Biology* **44**(5) (1998) 735.
[4] K.O. Greulich and G. Pilarczyk, Laser tweezers and optical microsurgery in cellular and molecular biology. Working principles and selected applications, *Cellular and Molecular Biology* **44**(5) (1998) 701.
[5] M.D. Wang, Manipulation of single molecules in biology, *Current Opinion in Biotechnology*, **10**(1) (1999) 81.
[6] S.C. Kuo, Using optics to measure biological forces and mechanics, *Traffic* **2**(11) (2001) 757.
[7] G.V. Shivashankar, Mesoscopic biology, *Pramana-Journal of Physics* **58**(2) (2002) 439.
[8] P.T. Korda, G.C. Spalding, and D.G. Grier, Evolution of a colloidal critical state in an optical pinning potential landscape, *Physical Review B* **66**(2) (2002).
[9] A. Ashkin, Optical trapping and manipulation of neutral particles using lasers, *Proceedings of the National Academy of Sciences of the United States of America* **94**(10) (1997) 4853.
[10] A. Ashkin, History of optical trapping and manipulation of small-neutral particle atoms and molecules, *IEEE Journal of Selected Topics in Quantum Electronics* **6**(6) (2000) 841.

[11] D.G. Grier, Optical tweezers in colloid and interface science, *Current Opinion in Colloid and Interface Science* **2**(3) (1997) 264.
[12] H. Lowen, Colloidal soft matter under external control, *Journal of Physics-Condensed Matter* **13**(24) (2001) R415.
[13] E.R. Dufresne and D.G. Grier, Optical Tweezer Arrays and Optical Substrates Created with Diffractive Optical Elements, *Reviews of Scientific Instruments*, **69**(5) (1998) 1974.
[14] E.R. Dufresne, G.C. Spalding, M.T. Dearing, S.A. Sheets, and D.G. Grier, Computer-Generated Holographic Optical Tweezer Arrays, *Reviews of Scientific Instruments* **72**(3) (2001) 1810.
[15] J.E. Curtis, B.A. Koss, and D.G. Grier, Dynamic holographic optical tweezers, *Optical Communications* **207**(1-6) (2002) 169.
[16] J.E. Curtis and D.G. Grier, Structure of Optical Vortices, submitted to *Physical Review Letters*.
[17] A. Ashkin, Forces of a Single-Beam Gradient Laser Trap on a Dielectric Sphere in the Ray Optics Regime. In: M.P. Sheetz (ed.), *Laser Tweezers in Cell Biology*, Academic Press, 1998, pp 1-27.
[18] A.L. Stout and W.W. Webb, Optical Force Microscopy. In: M.P. Sheetz (ed.), *Laser Tweezers in Cell Biology*, Academic Press, 1998, pp 99-116.
[19] K. Helmerson, R. Kishore, W.D. Phillips, and H.H. Weetal, Optical tweezers-based immunosensor detects femtomolar concentrations of antigens, *Clinical Chemistry* **43**(2) (1997) 379.
[20] A.L. Stout, Detection ad Characterization of Individual Intermolecular Bonds Using Optical Tweezers, *Biophysical Journal* **80** (2001) 2976.
[21] S.B. Smith, Y. Cui, and C. Bustamante, Overstretching B-DNA: The Elastic Response of Individual Double-Stranded and Single-Stranded DNA Molecules, *Science* **271** (1996) 795.
[22] P. Cluzel, A. Lebrun, C. Heller, R. Lavery, J.L. Viovy, D. Chatenay, and F. Caron, DNA: An extensible molecule, *Science* **271** (1996) 792.
[23] C. Bustamante, S.B. Smith, J. Liphardt, and D. Smith, Single-molecule studies of DNA mechanics, *Current Opinion in Structural Biology* **10** (2000) 279.
[24] C. Bustamante, J.C. Macosko, and G.J.L. Wuite, Grabbing the Cat by the Tail: Manipulating Molecules One by One, *NatureReviews, Molecular Cell Biology* **1** (2000) 131.
[25] M.L. Bennink, L.H. Pope, S.H. Leuba, B.G. de Grooth, and J. Greve, Single Chromatin Fibre Assembly Using Optical Tweezers, *Single Molecules* **2** (2001) 91.
[26] M.L. Bennink, O.D. Scharer, R. Kanaar, K. Sakata-Sogawa, J.M. Schins, J.S. Kanger, B.G. de Grooth, and J. Greve, Single-Molecule Manipulation of Double-Stranded DNA Using Optical Tweezers: Interaction Studies of DNA with RecA and YOYO-1, *Cytometry* **36** (1999) 200.
[27] B.D. Brower-Toland, C.L. Smith, R.C. Yeh, J.T. Lis, C.L. Peterson, and M.D. Wang, Mechanical disruption of individual nucleosomes reveals a reversible multistage release of DNA, *Proceedings of the National Academy of Sciences of America* **99** (4) (2002) 1960.
[28] U. Bockelmann, P. Thomen, B. Essevaz-Roulet, V. Viasnoff, and F. Heslot, Unzipping DNA with optical tweezers: high sequence sensitivity and force flips, *Biophysical Journal* **82**(3) (2002) 1537.
[29] S.J. Koch, A. Shundrovsky, B.C. Jantzen, and M.D. Wang, Probing protein-DNA interactions by unzipping a single DNA double helix, *Biophysical Journal* **83**(2) (2002) 1098.
[30] K. Hirano, Y. Baba, Y. Matsuzawa, and A. Mizuno, Manipulation of single coiled DNA molecules by laser clustering of microparticles, *Applied Physics Letters* **80**(3) (2002) 515.
[31] A.D. Mehta, K.A. Pullen, and J.A. Spudich, Single molecule biochemistry using optical tweezers, *FEBS Letters* **430** (1998) 23.
[32] A.E. Knight, C. Veigel, C. Chambers, and J.E. Molloy, Analysis of single-molecule mechanical recordings: application to acto-myosin interactions, *Progress in Biophysics and Molecular Biology* **77** (2001) 45.
[33] A. Ishijima and T. Yanagida, Single Molecule Nanobioscience, *TRENDS in Biochemical Sciences* **26**(7) (2001) 438.
[34] C.K. Sun, Y.C. Huang, P.C. Cheng, H.C. Liu, and B.L. Lin, Cell manipulation by use of diamond microparticles as handles of optical tweezers, *Journal of the Optical Society of America B* **18**(10) (2001) 1483.
[35] Z.P. Luo, and K.N. An, Development and validation of a nanometer manipulation and measurement system for biomechanical testing of single macro-molecules, *Journal of Biomechanics* **31** (1998) 1075.
[36] M.C. Williams and I. Rouzina, Force Spectroscopy of single DNA and RNA molecules, *Current Opinion in Structural Biology* **12** (2002) 330.
[37] J. Wakayama, M. Shohara, C. Yagi, H. Ono, N. Miyake, Y. Kunioka, and T. Yamada, Zigzag motions of the myosin-coated beads actively sliding along actin filaments suspended between immobilized beads, *Biochimica et Biophysica Acta* **1573** (1999) 93.
[38] K. Kawaguchi and S. Ishiwata, Temperature Dependence of Force, Velocity, and Processivity of Single Kinesin Molecules, *Biochemical and Biophysical Research Communications* **272** (2000) 895.
[39] H. Felgner, R. Frank, and M. Schliwa, Flexural rigidity of microtubules measured with the use of optical tweezers, *Journal of Cell Science* **109** (1996) 509.

[40] H. Felgner, R. Frank, J. Biernat, E. Mandelkow, E. Mandelkow, B. Ludlin, A. Matus, and M. Schliwa, Domains of Neuronal Microtubule-associated Proteins and Flexural Rigidity of Microtubules, *Journal of Cell Biology* **138**(5) (1997) 1067.
[41] L. Finzi, P. Galajada, and G. Garab, Labeling phosphorylated LHCII with microspheres for tracking studies and force measurements, *Journal of Photochemistry and Photobiology B: Biology* **65** (2001) 1.
[42] J. Dai and M.P. Sheetz, Cell Membrane Mechanics. In: M.P. Sheetz (ed.), *Laser Tweezers in Cell Biology*, Academic Press, 1998, pp 157-171.
[43] A. Kusumi, Y. Sako, T. Fujiwara, and M. Tomishige, Application of Laser Tweezers to Studies of the Fences and Tethers of the Membrane Skeleton that Regulate the Movements of Plasma Membrane Proteins. In: M.P. Sheetz (ed.), *Laser Tweezers in Cell Biology, Academic Press*, 1998, pp 174-194.
[44] T. Fujii, Y.L. Sun, K.N. An, and Z.P. Luo, Mechanical properties of single hyaluronan molecules, *Journal of Biomechanics* **35** (2002) 527.
[45] A. Teray, J. Oakey, and D.W.M Marr, Fabrication of linear colloidal structures for microfluidic applications, *Applied Physics Letters* **81**(9) (2002) 1555.
[46] R. Omori and A. Suzuki, Uranium dioxide particles collection using radiation pressure of a laser light in air, *Journal of Nuclear Science and Technology* **35**(11) (1998) 830.
[47] R. Omori, K. Shima, and A. Suzuki, Rotation of optically trapped particles in air, *Japanese Journal of Applied Physics Part 2 – Letters* **38**(7A) (1999) L743.
[48] P. Korda, G.C. Spalding, E.R. Dufresne, and D.G. Grier, Nanofabrication with Holographic Optical Tweezers *Reviews of Scientific Instruments* **73** (2002) 1956.
[49] P.T. Korda and D.G. Grier, Annealing thin colloidal crystals with optical gradient forces, *Journal of Chemical Physics* **114**(17) (2001) 7570.
[50] P. Galajda and P. Ormos, Rotors produced and driven in laser tweezers with reversed direction of rotation, *Applied Physics Letters* **80**(24) (2002) 4653.
[51] P. Galajda and P. Ormos, Complex micromachines produced and driven by light, *Applied Physics Letters* **78**(2) (2001) 249.
[52] A. Ashkin and J.M. Dziedzic, Optical Trapping and Manipulation of Single Living Cells Using Infrared-Laser Beams, *Nature* **330**(6150) (1987) 769.
[53] A. Ashkin and J.M. Dziedzic, Optical Trapping and Manipulation of Viruses and Bacteria, *Science* **235**(4795) (1987) 1517.
[54] J. Fröhlich, H. König, New techniques for isolation of single prokaryotic cells, *FEMS Microbiology Reviews* **24** (2000) 567.
[55] M.W. Berns, Y. Tadir, H. Liang, and B. Tromberg, Laser Scissors and Tweezers. In: M.P. Sheetz (ed.), *Laser Tweezers in Cell Biology*, Academic Press, 1998, pp 71-98.
[56] Y. Tadir, W.H. Wright, O. Vafa, T. Ord, R.H. Asch, and M.W. Berns, Micromanipulation of sperm by a laser generated optical trap, *Fertility and Sterility* **52** (1989) 870.
[57] Y.L. Sun, Z.P. Luo, and K.N. An, Stretching Short Biopolymers Using Optical Tweezers, *Biochemical and Biophysical Research Communications* **286** (2001) 826.
[58] S.C. Grover, R.C. Gauthier, and A.G. Skirtach, Analysis of the behaviour of erythrocytes in an optical trapping system, *Optics Express* **7**(13) (2000) 533.
[59] J. Guck, R. Ananthakrishnan, C.C. Cunningham, and J. Kas, Stretching biological cells with light, *Journal of Physics-Condensed Matter* **14**(19) (2002) 4843.
[60] D.J. Odde and M.J. Renn, Laser-guided direct writing for applications in biotechnology, *TIBTECH* **17** (1999) 385.
[61] D.J. Odde and M.J. Renn, Laser-Guided Direct Writing of Living Cells, *Biotechnology and Bioengineering* **67** (2000) 312.
[62] H. Felgner, F. Grolig, O. Muller, and M. Schliwa, In Vivo Manipulation of Internal Cell Organelles. In: M.P. Sheetz (ed.), *Laser Tweezers in Cell Biology*, Academic Press, 1998, pp 195-203.
[63] M.A. Welte, S.P. Gross, M. Postner, S.M. Block, and E.F. Wieschaus, Developmental Regulation of Vesicle Transport in Drosophila Embryos: Force and Kinetics, *ScienceDirect* (2000).
[64] N. Ponelies, J. Scheef, A. Harim, G. Leitz, and K.O. Greulich, Laser Micromanipulators for Biotechnology and Genome Research, *Journal of Biotechnology* **35**(2-3) (1994) 109.
[65] S. Katsura, K. Hirano, Y. Matsuzawa, K. Yoshikawa, and A. Mizuno, *Nucleic Acids Research* **26**(21) (1998) 4943.
[66] S. Katsura, A. Yamaguchi, K. Hirano, Y. Matsuzawa, and A. Mizuno, Manipulation of globular DNA molecules for sizing and separation, *Electrophoresis* **21** (2000) 171.
[67] C.S. Buer, K.T. Gahagan, G.A. Swartzlander, and P.J. Weathers, Insertion of microscopic objects through plant cell walls using laser microsurgery, *Biotechnology and Bioengineering* **60** (1998) 348.
[68] S.C. Grover, A.G. Skirtach, R.C. Gauthier, and C.P. Grover, Automated single-cell sorting system based on optical trapping, *Journal of Biomedical Optics* **6**(1) (2001) 14.

[69] L. Brewer, M. Corzett, and R. Balhorn, Condensation of DNA by spermatid basic nuclear proteins, *Journal of Biological Chemistry* **277** (41) (2002) 38895.
[70] Zahn, 1999.
[71] P.T. Korda, M.B. Taylor, and D.G. Grier, Kinetically locked-in colloidal transport in an array of optical tweezers, *Physical Review Letters* **89**(12) (2002).
[72] A. Ehrlicher, T. Betz, B. Stuhrmann, D. Koch, V. Milner, M.G. Raizen, and J. Kas, Guiding neuronal growth with light, *Proceedings of the National Academy of Sciences of the United States of America* **99**(25) (2002) 16024.
[73] K.A. Blanks, Novel synthesis of gibbsite by laser-stimulated nucleation in supersaturated sodium aluminate solutions, *Journal of Crystal Growth* **220** (2000) 572.
[74] B.A. Garetz and J. Matic, Polarization Switching of Crystal Structure in the Nonphotochemical Light-induced Nucleation of Supersaturated Aqueous Glycine Solutions, *Physical Review Letters* **89**(17) (2002) .
[75] P.A. Bancel, V.B. Cajipe, F. Rodier, and J. Witz, Laser seeding for biomolecular crystallization, *Journal of Crystal Growth* **191** (1998) 537.
[76] P.A.Bancel, V.B. Cajipe, and F. Rodier, Manipulating crystals with light, *Journal of Crystal Growth* **196** (1999) 685.
[77] K. Ajito and M. Morita, Imaging and spectroscopic analysis of single microdroplets containing *p*-cresol using the near-infrared laser tweezers/Raman microprobe system, *Surface Science* **427-428** (1999) 141.
[78] J.P. Delville, C. Llaude, and A. Ducasse, Kinetics of laser-driven phase separation induced by a tightly focused wave in binary liquid mixtures, *Physica A* **262** (1999) 40.
[79] A. Pralle, E.-L Florin, E.H.K. Stelzer, and J.K.H. Hörber, Local viscosity probed by photonic force microscopy, *Applied Physics A*. **66** (1998) S71.
[80] R. Lugowski, B. Kolodziejczyk, and Y. Kawata, Application of laser-trapping technique for measuring the three-dimensional distribution of viscosity, *Optics Communications* **202**(1-3) (2002) 1.
[81] A. Lachish-Zalait, D. Zbaida, E. Klein, and M. Elbaum, Direct surface patterning from solutions: Localized microchemistry using a focused laser, *Advanced Functional Materials* **11**(3) (2001) 218.
[82] Y.K. Nahmias and D.J. Odde, Analysis of radiation forces in laser trapping and laser-guided direct writing applications, *IEEE Journal of Quantum Electronics* **38**(2) (2002) 131.
[83] P. Kral, and H.R. Sadeghpour, Laser spinning of nanotubes: A path to fast-rotating microdevices, *Physical Review B* **65**(16) (2002).
[84] K.D. Bonin, B. Kourmanov, and T.G. Walker, Light torque nanocontrol, nanomotors, and nanorockers, *Optics Express* **10** (19) (2002) 984.
[85] S. Katsura, A. Yamaguchi, H. Inami, S. Matsuura, K. Hirano, and A. Mizuno, Indirect micromanipulation of single molecules in water-in-oil emulsion, *Electrophoresis* **22**(2) (2001) 289.

Subject Index

Author Index